REGENERATING THE COALFIELDS

For Carlo

Regenerating the Coalfields

Policy and politics in the 1980s and early 1990s

ROYCE TURNER

Policy Research Centre
Sheffield Business School
Sheffield Hallam University

Avebury

Aldershot · Brookfield USA · Hong Kong · Singapore · Sydney

© Royce Turner 1993

Published by
Avebury
Ashgate Publishing Limited
Gower House
Croft Road
Aldershot
Hants GU11 3HR
England

Ashgate Publishing Company
Old Post Road
Brookfield
Vermont 05036
USA

British Library Cataloguing in Publication Data

Turner, Royce Logan
    Regenerating the Coalfields: Policy and Politics
    in the 1980s and Early 1990s
    I. Title
    338.27240941

ISBN 1 85628 414 X

Printed and Bound in Great Britain by
Athenaeum Press Ltd, Newcastle upon Tyne.

Contents

Figures and tables   ix

Acknowledgements   x

Preface   xi

Abbreviations   xiii

1   Introduction   **Page** **1**

1.1   Scope, objectives and method   1
1.2   Objectives of economic regeneration   8
1.3   National and regional economies   9
1.4   The role of politics   13
1.5   Defining the coalfields   14
1.6   Deindustrialization   20
1.7   Reindustrialization and the 'State'   25
1.8   Post-coal regeneration: factors
         justifying special attention   26

2   **Policy agenda determination, the**
         coalfields and the Coalfields
         Communities Campaign   **Page 35**

2.1   Introduction: agendas and demands on
        governments   35
2.2   Models of agenda building   38
2.3   Agenda building and the coalfields   40
2.4   Coalfield Communities Campaign   43
2.5   Legitimacy   46
2.6   Feasibility   48
2.7   Support   50
2.8   Additionality   58
2.9   Power and pressure groups   63
2.10   Conclusion   72

3   Policy making for regeneration in the
        coalfields: a changing framework
                                           Page 75
3.1   From regional policy to local
        projects   75
3.2   Policy communities and policy
        networks   80
3.3   Conclusion   87

4   Deindustrialization and
        reindustrialization in Bolsover:
        a multi-agency response to economic
        change in the North Derbyshire
        coalfield   Page 89

4.1   The locality   89
4.2   Deindustrialization   93
4.3   Regeneration   99
4.4   Bolsover Enterprise Park   108
4.5   The 'new' entrepreneur   110
4.6   Conclusion   114

5   British Coal Enterprise: bringing the
        'enterprise culture' to a
        deindustrialized local economy? Page 119

5.1   Introduction   119
5.2   Coal: the scale of
        deindustrialization   121
5.3   The locality   121
5.4   BCE operations at Carcroft   122
5.5   The enterprise culture   124
5.6   BCE and evaluation   128
5.7   The outcome   130
5.8   Conclusion   133

6   An enterprise zone in a pit closure
         zone: the politics of industrial
         subsidies   Page 136

6.1   A 'politics' of industrial
         subsidies?   136
6.2   The location   137
6.3   The enterprise zone idea   139
6.4   Reality and analysis   142
6.5   Conclusion   159

7   A Task Force in a locality of coal
         mining decline: the case of Doncaster
                                          Page 161
7.1   Introduction   161
7.2   The Task Force idea   163
7.3   Objectives   165
7.4   The locality   172
7.5   The resources   174
7.6   Doncaster Chamber of Commerce
         and Industry   176
7.7   Training   178
7.8   Conclusions   183

8   Barnsley Business and Innovation
         Centre: the impact of a strategy for
         innovation   Page 185

8.1   The idea   185
8.2   Technological modernisation and
         government intervention   190
8.3   The structure of innovation   199
         centres
8.4   Barnsley Business and Innovation
         Centre   201
8.5   Locality and local economy   205
8.6   Budgets and agendas   206
8.7   Objectives and criteria for
         assessment of projects   210
8.8   Impacts   213
8.9   Conclusions   220

9   Conclusions   Page 228

Appendix 1:   Postal questionnaire to companies
               on Langthwaite Grange Industrial
               Estate, South Kirkby.   Carried
               out September to December 1990
                                               236

Appendix 2:   Postal questionnaire to selected
                 companies   in   Barnsley   on
                 employment   structure   and
                 characteristics.   Carried out
                 February 1992
                                                239

Appendix 3:   Local   Authority   members   of
                 Coalfield Communities Campaign as
                 of April 1991
                                                242

Appendix 4:   Business rates applicable in City
                 of   Wakefield   Metropolitan
                 District Council administrative
                 area 1981/2 to 1990/91
                                                245

Appendix 5:   Analysis of subsidies paid to
                 South   Kirkby   enterprise   zone
                 companies
                                                246
References   251

Figure 1   Location of collieries within 5
                  miles of South Kirkby   154

Figure 2   Doncaster Task Force boundaries
                                                 168

Table 1   Coal Mines within 5 miles of
                  Bolsover town 1968 onwards
                                                  92

Table 2   Businesses and paid employment on
                  Bolsover Enterprise Park as of
                  July 1991   103-104

Table 3   Businesses on Bolsover Enterprise
                  Park and their responses to
                  business advice from Rural
                  Development Commission and
                  British Coal Enterprise
                                             117-118

Table 4   Actual and expected take-up of
                  Enterprise Allowance Scheme 1985
                  in the 'Five Towns'   128

Table 5   Categorisation of 25 firms at
                  Carcroft Enterprise Park as of
                  February 1991   132

Table 6   UK Business and Innovation
                  Centres that were also members of
                  the European Business and
                  Innovation Centre Network as of
                  March 1992   187

Table 7   Collieries and associated works
                  closed within a 5 mile radius of
                  Barnsley Business and Innovation
                  Centre   208

Table 8   Barnsley Business and Innovation
                  Centre. Summary of Findings (a)
                  April 1992   226

Table 9   Barnsley Business and Innovation
                  Centre. Summary of Findings (b)
                  April 1992   227

Acknowledgements

Many people gave generously of their time and contributed information invaluable in the composing of this work. One hundred and twenty-nine people, for example, were interviewed. They are therefore too numerous to name individually, and some of them would wish anonymity anyway. I would, however, like especially to thank Barbara Edwards, of the Coalfield Communities Campaign; Peter Stafford, of Bolsover Enterprise Agency; Sue Linsley, of Derbyshire County Council Planning Department; Joe Armishaw, of British Coal Enterprise; David Twigg, of Doncaster Task Force; and Dr Brian King, of Barnsley Business and Innovation Centre. None of the above were responsible for the interpretation of the data presented here. Some, if not all, might have disagreed with it. The responsibility for the interpretation lies with the author. I would also like to thank Helen Escott, Jilly Cooper, Linda Bilson and Ellenor Moody for a tremendous job in typing a difficult manuscript.

Preface

In October 1992 the politics of de-industrialization appeared, at least momentarily, to have changed. British Coal announced the closure or moth-balling of 31 of its 50 collieries and the shedding of 30,000 jobs. It was to be the closure, at a stroke, of deep-mining in places which had been synonymous with it: it would end in Doncaster; in Lancashire; it would virtually end in the north east; in Barnsley; in South Wales; in Scotland. Public opinion erupted in anger. There was widespread sympathy for the mining communities. Newspapers usually loyal to the Conservative Party hit out against the Conservative government and called upon them to change policy. There was outrage across the country. Public opinion achieved what pressure groups had failed to achieve: it placed the issue of pit closures, and the problems caused by pit closures, on the political agenda for the first time. The government made some efforts to modify policy on closures following this.

The work was completed prior to the October 1992 announcement. The argument in Chapter two, therefore, should be read as an

account and analysis of the political situation
prior to October 1992.

One factor however was made abundantly
clear in October 1992: the need for economic
regeneration in the coalfields became more
pressing than ever before. Judging by the
projects examined in this work, those economic
regeneration efforts would have to be intensive
and extensive if they were going to make
anything more than a modest impact.

Abbreviations

BACM	British Association of Colliery Managers
BBIC	Barnsley Business and Innovation Centre
BCE	British Coal Enterprise
CCC	Coalfield Communities Campaign
CEC	Commission of the European Communities
CEGB	Central Electricity Generating Board
DTI	Department of Trade and Industry
EETPU	Electrical Electronic Telecommunication and Plumbing Union
FOIL	Fight Opencast in Leicestershire
MNC	Multinational Corporation
NACRO	National Association for the Care and Resettlement of Offenders

NCB	National Coal Board
NUM	National Union of Mineworkers
OECD	Organisation for Economic Co-operation and Development
RDC	Rural Development Commission
RECHAR	Restructuration des Charbonnages. A fund of money available from the European Community for areas experiencing coal mining decline
TUC	Trade Union Congress

1 Introduction

1.1 Scope, objectives and method

This work is about economic change in what have been, in the post-Second World War era, territorially, some of Britain's major coalfield localities. The central objective of the work was to examine and evaluate a sample of the economic regeneration strategies that have been adopted in some of the relevant localities.

The scale of deindustrialization in the coalfields has been vast, and was effected rapidly, in the late 1980s and early 1990s. Of the 191 National Coal Board (NCB) collieries operating at the end of the 1983 financial year, employing 207,600 (NCB, 1984), by July 1992 there were only 50 pits being operated by the renamed British Coal, employing 58,100.

Most of the economic regeneration strategies adopted in the coalfields had a governmental or quasi-governmental input at the level of local or central government; some had a voluntary sector (Young, 1985), or private sector input, or both.

This work is concerned with: why were particular policies/strategies for economic regeneration adopted in particular localities?

Did the strategies 'work'? In other words, did they provide jobs and/or economic activity sufficient to replace the jobs lost, and to counter the reduction in economic activity, occasioned by the decline in the deep coal mining industry in the late 1980s and early 1990s? If so, at what cost to the public purse? In short, did they succeed in 'regenerating' the economies within the targeted local territories?

Such an investigation necessitated the examination and evaluation of a spectrum of regeneration initiatives. In some cases, these were central government-inspired initiatives, such as the enterprise zone development commenced in 1981 in South Kirkby in West Yorkshire, or the Task Force which operated between 1987 and 1990 in Doncaster. On a larger scale, another example, though not examined here, of a central government-inspired regeneration initiative, was the Valley Regeneration Towns Project in South Wales. This was launched in 1988 as a central government-initiated and co-ordinated regeneration strategy, which also encompassed activity by local authorities, the Welsh Development Agency, voluntary associations, and the private sector (Welsh Office, undated; Welsh Office Information Division, 1988). The Valley Regeneration Towns Project was itself a successor to The Valleys Initiative, launched by central Government in February 1986 (Romaya and Alden, 1987 and 1988).

In other cases, it was local authorities that took the lead in spearheading economic regeneration initiatives. Some local authorities joined forces with other bodies to pursue the regeneration process. Bolsover District Council, for example, entered into the North Derbyshire Partnership (discussed in more depth later) in the early 1990s with British Coal Enterprise (BCE), the job creation subsidiary of British Coal, and local authorities bordering Bolsover or, in the case of one of the partners, Derbyshire County Council, having some administrative jurisdiction over Bolsover.

Barnsley Metropolitan Borough Council was another example. They entered into a partnership

2

with the construction company Costain in 1991
with the initial objective of providing new
office space, (Barnsley Chronicle, 4 October
1991), but also wider and much more ambitious
objectives:

'(i)   [the] .... creation of 10,000 good
        quality jobs;

(ii)   [to] ..create high quality
        development on major sites in the M1
        corridor and throughout the borough;

(iii)   [to] .... regenerate and improve the
        quality of life throughout the
        borough;

(iv)   the promotion of Barnsley nationally
        and locally,' (Barnsley Development
        Office, 1992).

The partnership was scheduled to last for 'a
minimum of 10 years' and would also 'carry
out/arrange associated training'. Barnsley
Metropolitan Borough Council and Costain each
had '50 per cent of the share capital and equal
voting rights' (Barnsley Development Office,
1992). The local authority invested £1.5
million in the project collected from community
charge payers.

Another important development in the
mid-1980s was the banding together of local
authorities which had had some connection with
the coal mining industry to form a multi-local
authority pressure group: the Coalfield
Communities Campaign. The objectives of this
pressure group were to fight against further
contraction in the coal industry, and to fight
for the adoption of economic regeneration
schemes and projects, backed by national
government and/or the European Community, in
localities suffering pit closures and associated
deindustrialization. In the context of this
work on policy making and implementation, the
efforts of the Coalfield Communities Campaign
are examined as a mechanism for attempting to
gain a recognition of the needs/demands/wants of
some of the people living in the coalfields on

3

the policy 'agenda'. This concept is examined in much greater depth in Chapter Two.

This work is not meant to be a manual 'of how to regenerate the coalfields'. Given that the precise nature of the problems preventing **spontaneous** economic regeneration in localities that were formerly involved in coal mining varied — access to, or remoteness from, large-scale markets for goods and services; quality of, and accessibility to, infrastructure; the nature of the local skills base — no such overall document could exist with general applicability. A document close to fitting the bill for a specific locality over a specific time period, however, might be the *Dearne Valley Initiative Economic Strategy Study and Business Plan*. This was published in 1990, and was commissioned by Barnsley, Doncaster and Rotherham Metropolitan Councils. Each of these authorities had administrative jurisdiction over parts of the Dearne Valley in South Yorkshire, and the document examined the Valley's strengths and weaknesses in relation to the potential for economic regeneration in a post-coal industry economy (Coopers and Lybrand Deloitte and Sheffield City Polytechnic, 1990). The Dearne Valley and its bid for assistance from central government is also discussed in Chapter Two.

If this study is not a consultant's report on 'how to regenerate', nor are there any hypotheses to test in a work of this sort. Though there are obviously disagreements as to why national and local economies deindustrialize, with the focus of debate often being on whether industrial decline has been a function of a failure of government economic policy or the fault of poor performance by companies themselves, or promoted by an anti-enterprise, anti-industrialist culture (see, for example, Blackaby, 1979; Pollard, 1982; Wiener, 1981; Williams et al 1983); there is broad agreement as to the outcome of the process. Markets for the output of national or local production units decline or disappear, or production is switched to other localities, sometimes overseas. Similarly, the rationales behind particular attempts at economic regeneration also do not provide a mystery,

4

where supposition is called for in an attempted explanation, and therefore a test of that supposition becomes necessary. That is not, of course, to argue that those rationales are always coherent and valid, or are not susceptible to question, as if they had emerged from some ideologically value-neutral system of 'economics'. If that were the case there would be no need for this work: policy analysis, economic policy evaluation, like large parts of the coal industry in the 1980s and 1990s, would have become redundant.

Rather, then, this work is essentially an exercise in policy analysis: how and why particular economic regeneration policies were embarked upon, and how successful these were.

That is not to state, though, that theory is unimportant to this thesis; it is important and the intention has been to fuse and conflate it into the empirical work in order to give substance to the latter. Therefore in the course of this work, theories on the determination of governing authorities' policy agendas are examined, for regeneration policy needs to be on an agenda if it is to be given policy attention. Similarly, theories on 'positive discrimination' in regeneration are examined; the theoretical justification, or lack of it, for industrial subsidies; the value of the promotion of entrepreneurship; and the theoretical justification for a governmental input into an attempt to secure a 'modernised' high technology-based local or national economy, for example. Chapters 2 and 3, in particular, examine theories of policy initiation, formulation and implementation. The remainder of the chapters consist more broadly of policy evaluation and analysis appropriate to the regeneration scheme being examined.

Methodologically, the central tools used for the collection of material for this study — and to help towards the evaluative analysis — were interviews with key actors in the economic regeneration agencies, interviews with owners or senior managers in companies in coalfield localities, and liaison with local authority planning and economic development departments. Details of who was interviewed and when are in the individual chapters. An attempt was made to employ a questionnaire (see Appendix 1) directed at companies on an enterprise zone in South Kirkby in West Yorkshire in 1990, and though the information collected was useful, the response rate was poor. Of the 75 questionnaires distributed in October 1990, just 25 were returned: a response rate of 30 per cent.

A further attempt at a postal questionnaire was made in March 1992. Eleven companies in the Barnsley region were contacted: Mercedes Benz (distribution); Dunlop Slazenger (tennis balls); SR Gent Plc (textiles); Midland Bank Computer Centre; Union Chemical of Japan (ink ribbon manufacture); Kostal UK (switchgear assembly for car manufactures); BOC (food storage and distribution); Maplin Electronics; Mydrin (Chemicals); Koyo Seiko of Japan (ballbearing manufacture); Lyons Bakeries. All these companies were identified by Barnsley Metropolitan Borough Council as having 'made the Barnsley choice', (Barnsley Metropolitan Borough Council, undated, circa 1991), and therefore, presumably, were seen as being indicative of the newly-regenerating post-coal industrial economy in the locality. The objective of this questionnaire was to provide a profile of the labour forces being employed by these companies: particularly in relation to whether or not they were employing people formerly employed in the coal industry (see Appendix 2). Again, however, the response rate was poor: only three companies replied. Two of those companies which responded — Midland Bank Computer Centre and Union Chemical of Japan — were at the time not employing anyone who had previously worked in the coal industry. The third respondent — BOC — did not keep records. The information received from this questionnaire was therefore not really a great deal of use.

Primary information gathering within this work was therefore mainly carried out by face-to-face or telephone interviews, and occasionally, where a company would respond only to written communication, or where a point needed verifying, by letter to companies, or to agencies engaged in economic regeneration.

In addition, of course, the body of literature that existed on government policy towards the coal industry, and on the politics and effectiveness of economic regeneration measures, was called upon. At the time of writing, however, there was little on the effectiveness or otherwise, or the politics, of regeneration measures in localities specifically associated with the coal industry (though see Hudson and Sadler, 1987; Beynon, Hudson and Sadler, 1991, pp.110-121; McNulty, 1987). Therefore there is no discrete 'literature review' within this work: an appreciation of published work which has a connection with the major subject matter here, rather, is conflated with the general evaluation and analysis within.

The evaluation methodology employed in this work varied in each case study, because the activities under scrutiny in each case study were different. In some instances, the design of the search was to discover whether or not 'new' entrepreneurship — entrepreneurship by people not previously engaged in it, and with no personal 'tradition' of entrepreneurship — had been stimulated by the activities of regeneration agencies. In another instance here, the cost to the public purse of subsidising employment in a Conservative government-sanctioned enterprise zone is compared to the cost per job of subsidising employment in a particular, named, coal mine. The uniting theme is whether or not jobs associated with a particular project were in fact 'new'— what Martin (1989) called 'attributable jobs': that is, attributable to the governmental or quasi-governmental assistance or stimulation — or were simply transferred from another locality or even from within the same locality. Another facet of this related to whether or not jobs already existing would have continued to exist, or whether 'new' jobs created would have been created anyway, in the absence of any activity by economic regeneration agencies. Determining this, of course, is difficult. Sometimes it is impossible to tell.

As Martin (1989) noted:

7

'In many studies therefore [after assessing other means of counting 'new' jobs] the only available approach has been to rely on information provided by the managing directors of firms which have received assistance. However, the perceptions of this group may not be entirely accurate....'

And even if the perceptions are accurate, the message conveyed might not be. Any company receiving financial or other assistance is unlikely to convey the message that such assistance is unnecessary to their survival or to their creation/maintenance of jobs: to do so would be to invite the curtailment of the assistance.

Whether or not jobs created were genuinely 'new', were transferred from near or far, or would have existed anyway, mattered: a musical chairs of transferred and transferring economic activity would not lead to the revivifying of regional economies; it may not lead to the revivifying of local economies, except at the expense of another.

The evaluative approach here attempts to make a common sense appraisal of whether or not jobs would have been created anyway in the locality under investigation based on, firstly, information from the companies themselves; and secondly, other salient evidence readily available. For example, a company providing services/commodities for/to the coal industry would hardly be located elsewhere other than in the coalfields.

1.2 Objectives of economic regeneration

Aside from whether or not jobs 'created' were actually 'new' jobs, another connected important question related to the following: how was successful economic regeneration in the coalfields to be recognised where it did exist? What would have constituted it?

The answer might appear obvious, but in fact is not. Essentially the answer depends on what the **objectives** of the particular economic regeneration scheme under investigation were. If the objective of an economic regeneration scheme was the creation of 'new' jobs then this is what would be tested in an evaluation; if it was the creation of 'new entrepreneurship', or 'innovation', then these are what would be looked for. Or if 'leverage' was the objective — the spending of some public money in order to initiate more spending by the private sector — then this is what would be tested for. Clearly, then, the 'performance indicators' (see Martin, 1990) used to evaluate the extent of policy success would vary from project to project.

1.3 National and regional economies

To an extent, economic regeneration in the coalfields in the 1980s and 1990s was predicated upon economic regeneration at the national level. It is recognised here that some of the prospects for the former are subsumed within the prospects for the latter. That is not to argue, however, that the prospects for regions and sub-regions within national economies are ineluctably and exclusively tied to the prospects of a national economy as a whole. That is not the case, and if it were, no region, sub-region or locality within a national economy would ever be more or less prosperous than another region, sub-region or locality. Sabel (1989) has argued that since the mid-1970s some of the advanced national economies have seen a re-emergence of the 'regional economy,' if not as an autonomous entity, at least as an identifiable part of a national economy within given territorial locations, within the boundaries of which companies and economically-active individuals can be found relating to, and inter-acting with, each other. Sometimes these companies would make similar products; sometimes they would act as sub-contractors for each other or draw on a common body of sub-contractors; sometimes they would use each other's capacity; sometimes companies would have representatives sitting on the boards of co-operative regional banks authorising loans for other companies in this 'economic community'. As such, the 'regional economy' could have a coherent identity which separated it from the national economy.

According to Sabel, this re-emergence of the regional economy had several notable territorial manifestations, where the economy of the region or sub-region stood out as being 'different' from the surrounding economy. He identified 'Third Italy,' which was distinct from the old industrial centres of Genoa, Turin and Milan, and distinct from the impoverished south of the country. Rather:

> 'It is a string of industrial districts stretching from the Venetian provinces in the North through Bologna and Florence to Ancona in the south, and producing every thing from knitted goods (Capri), to special machines (Parma, Bologna), ceramic tiles (Sassuolo), textiles (Como, Prato), agricultural implements (Reggio Emilia), hydraulic devices (Modena), shoes, white goods, plastic tableware, and electronic musical instruments (Ancona)' (Sabel, 1989, p.22).

There were other areas which Sabel characterised as being 'new regional economies,' which were distinct from their wider national economies because of the high level of industrial activity that was obtaining within them. 'Second Denmark' was identified as being, in the late 1980s:

> 'A patchwork of textile, garment, furniture, machine-tool, and shipbuilding districts that are outgrowing the established centres of industry surrounding Copenhagen in the East' (Sabel, 1989, p.22).

Similarly identified were the 'Swedish metalworking producers in Småland,' industrial districts in the **Land** of Baden-Württemberg in West Germany, and 'semi-conductor production in Silicon Valley, south-east of San Francisco, and the concentration of mini-computer producers along Route 128 circling Boston.'

What distinguished these 'new regional economies' in the 1980s was that, firstly, they outperformed economically other areas elsewhere in the respective country. Secondly, they had developed economic regimes of 'flexible specialisation' as opposed to mass production. Mass production has also been termed 'Fordism', after the vehicle manufacture which was one of its earliest advocates and with which it is probably most closely associated. 'Fordism,' essentially, involved the mass production of commodities which were always almost the same as each other; assembly line production by serried ranks of workers each engaged in a specific, routine and repetitive task designed on the 'scientific management' principles of FW Taylor; continuous production; massive investment in capital. The corollary of mass production was mass consumption. Very often, this was mass consumption of the same or similar products: the Ford car; the McDonalds hamburger. With the rise of the multinational corporation, mass production Fordism was exported worldwide. The commodity which perhaps best epitomised its rise was the attempt to design a 'world car' — which could be built and driven in virtually any part of the world — in the 1970s and 1980s by giant corporations such as General Motors or Ford.

Flexible specialisation — or neo-Fordism — on the other hand, was characterised by the abandonment of the rigid assembly-line production method. Changes in technology meant that smaller batch production became economically viable. Computer-aided design contributed to this, and computer-management of stocks meant that less inventory needed be held: just-in-time replaced the necessity of having vast stocks of materials lying idle. Production was also characterised by the rise of team work; the increasing use of sub-contractors and workers/ managers/consultants outside the corporation/ company in question; decentralisation in firm organisation, so that small, autonomous units were established to handle the production of one range of commodities within the portfolio of the company as a whole; and a decline in large- scale unionism, (Harris, 1988). Flexible specialisation allowed for the development of a more widely differentiated product range, which could be altered more frequently than under mass production. The net result was that, at least in theory, the consumer should benefit from greater product choice.

A third factor characterising the 'new regional economies' was that they **were** geographically-distinct and territorially-circumscribed. In this sense, theoretically, they would have something in common with coalfields which are or have been also geographically-distinct and territorially-circumscribed. A fourth factor characterising 'new regional economies' was that sometimes, though apparently not always, companies were producing similar goods or services, and there was an element of non-competitive co-operation between them (Sabel, 1989).

The purpose of this excursion into the arguments suggesting the emergence of 'new regional economies' was to demonstrate that, theoretically, it **was** possible for a regional or sub-regional economy to become relatively prosperous, even if a national economy was performing less well than had been hoped or expected. Therefore, at least theoretically, it would be possible for coalfield or former coalfield economies to undergo some level of economic regeneration even if the national economy as a whole was underperforming. It was not intended to infer that this was a likelihood, however.

Certainly it was the case in Britain, for at least parts of the 1980s and for the early 1990s, that the national economy was performing less well than had been hoped for, or expected, by many. Although the exploration of why that might have been the case — and its more long-standing 'relative economic decline' — is a worthy debate, it is too wide a canvass upon which to base this study. In any case, the debate has already been engaged in extensively by others (see, for example, Coates and Hillard, 1986; Gamble, 1981; Hirst and Zeitlin, 1989; Williams et al, 1983).

The focus within this study therefore remains for the most part narrow: it concentrates on selected coalfields/former coalfields and the regeneration agencies, firms and other relevant

12

factors within them. Though the condition of the national economy in Britain in the 1980s and early 1990s was important, it remained a backcloth against which the workings of other economic factors more direct to the coalfields could be observed.

1.4 The role of politics

A central recognition in this work is that decision-making, policy implementation and, indeed, policy evaluation takes place in a political environment. Policy is about politics (Carley, 1989,p.21).

There is no value-neutral quasi-science of public adminstration or economics where governmental decisions simply represent the interactions of ideology-free economic variables, or the concomitant actions of disinterested economic and/or political actors.

Decisions on pit closures in the first place, for example, were taken ostensibly on the grounds of 'economics', but economics is always coloured by politics (Alt and Chrystal, 1983). For instance a governing party might decide to broaden the means available of generating electricity — for example, by the development of nuclear power, hydroelectric generating stations, the importation of primary energy — in order to diminish the potential and/or actual power of a sectional interest group whose members control the production/extraction of a particular primary energy commodity such as coal. This indeed was what laid behind some of the recommendations of the Ridley Report on nationalised industries, prepared while the Conservative Party were in opposition in the 1970s, and leaked to The Economist magazine in 1978. Nicholas Ridley would later serve in Mrs Thatcher's cabinet, and could, without fear of contradiction, be classed as an ideological soulmate of Mrs Thatcher's (see, for example, Ridley 1991).

As mechanisms for countering any 'political threat' from 'the enemies of the next Tory Government,' the Ridley Report recommended, inter alia, the introduction of more dual oil/coal firing in electricity generating stations, and contingency plans for the importation of coal (Scammell, 1986).

There is clearly a strong 'political' element in reasoning of this kind, **as well** as an economic justification for broadening the base of electricity production. Relying largely on one energy source, produced by a monopoly supplier of labour, for the production of electricity, could cogently be argued as being 'bad' economics. The reduction of the power of a vested, sectional interest group, if such a move could be portrayed as being part of a move away from 'bad' economics, might also just happen to be politically convenient.

There are many other examples of economics being coloured by politics: the debate as to whether business/industry should be exclusively privately-owned, or whether there was scope for public sector ownership of it, owed as much to ideology as it did to 'technical' economics. The same point was applicable to almost any area of macro or micro economic policy: interest rate levels; direct and indirect taxation levels; levels of transfer payments (unemployment pay, pensions); regional policy, or its absence; financial assistance to industry, or its absence, and so on. The arguments were political arguments informed by economics, rather than being a product of a value-neutral science called economics.

1.5 Defining the coalfields

It is important at the outset to determine a geographical and economic definition of what actually constituted a 'coalfield'. This is for at least two reasons.

Firstly, there were competing definitions of what level of economic activity, in a locality or area, was necessary before that locality or area could be properly designated a 'coalfield'.

As a starting point for a definition, Fothergill and Gudgin (1985) noted, for example, that in 1985, two thirds of the total employment in coal mining was within a 70 mile swathe of land from Leeds to Nottingham, and within the adjacent but small coalfields in the

Leicestershire, South Derbyshire and West Midlands coalfields.

A more precise definition was formulated by the Commission of the European Communities (CEC). This was by way of determining eligibility criteria for monies from its RECHAR initiative launched in December 1989 to provide assistance to areas suffering coal mining decline. To qualify, administrative areas would have to meet the following:

'a)   contain, or have contained, deep or opencast mines for the extraction of hard coal or black lignite;

b)   must have employed at least 1,000 people in coal extraction on January 1st 1984 or later; and

c)   have lost at least 1,000 coal mining jobs since January 1st 1984 or the total number of coal mining jobs lost and at risk since January 1st 1984 must be at least 1,000.' (LEDIS, March 1990, emphasis in text).

Definitions are important. On the above, local authorities, companies and individuals in the west County Durham coalfield, for example, where coalmining had ceased by 1981 (Fothergill and Gudgin, 1985), would not be eligible for assistance from RECHAR. Being more specific, companies and individuals within the administrative territory of the Wear Valley local authority in County Durham, where West Auckland, Brancepeth, Stanley Cottage and Hole in the Wall? Crook Drift collieries all closed in the 1960s, would not meet the criteria necessary to gain assistance under RECHAR. The importance of definitions is also evident in the reverse sense to the above. British Coal Enterprise (BCE), for example, has been criticised for operating over too wide an area, offering assistance to companies in local economies where coal mining contraction has had little significance (Owen, 1988, p.6). Here would be an example of a too wide a definition of a coalfield.

15

The second reason for this focus on the geographical and economic definition of what actually constituted a 'coalfield' related to the extractive nature of the industry, and the finite nature of the commodity itself that was being mined. The coal industry was not a static industry; it moved from the extraction of one lot of coal deposits to the next. Any attempt at policy analysis here therefore has to have a temporal, as well as a spatial, definition.

For example, the large-scale extraction of coal reserves in the Vale of York — with the giant Selby coalfield project — was a new departure in the 1970s (National Coal Board, 1975).

Elsewhere coal mining has ceased as an industrial activity. Deep-mining ceased altogether in the Somerset coalfield, for example, with the closure of the Kilmersdon colliery near Radstock in 1973 (*The Times*, 31 May 1973).

It should be noted here, however, that the doppelganger of deep coal mining, in the far more environmentally hostile form of opencasting, could make a visitation long after the colliery winding gear had disappeared. 1991 saw British Coal announce it intended to prospect for opencast coal on three sites north of Bristol, in the old Somerset coalfield, for example (*The Guardian*, 11 October 1991). Indeed, in the late 1980s and early 1990s, the open cast doppleganger began to revisit many localities that had seen closures in, and sometimes the end of, their deep-mined coal industry. Localities such as South Cumnock, Ayr Valley, Forth, Monklands, Motherwell and the Doon Valley in Scotland, for example (Parkinson, 1985); the Gawber/Mapplewell localities (*Barnsley Chronicle*, 8 November 1991), and on the sites of the former Barrow and Rockingham collieries in Barnsley; the Normanton, Heather, Rathenstone and Packington localities in Leicestershire, some of the people of which formed the pressure group, FOIL, (Fight Opencast in Leicestershire) in 1989 to try to stop the activities (*The Guardian*, 11 October 1991). In County Durham, opencast coal production rose from 572,000 tonnes in 1975/6 to 1,004,000 tonnes in 1984/5 (Beynon, Hudson and Sadler, 1991).

This work is concerned primarily with the late 1980s and early 1990s. The reason for this is obvious: it would have been impossible, from the perspective of the early 1990s, to gain anything other than background information on economic regeneration in former mining areas which faced pit closures in the 1950s and 1960s, for example. These areas could be examined to see if patterns of, for instance, migration of former miners had replicated themselves, or if economic problems still lingered from the time of pit closures. But little else could be gained, given the extent of changes in individual local economies, and in the national economy, since then. In the 1960s, for instance, near full employment obtained in most regions. In the 1980s and early 1990s, it did not. The 1960s also saw growth in some industries, such as vehicle manufacture, electrical engineering, chemicals and petroleum products (Martin, 1988), even if others, such as coal, were in decline. The 1980s and early 1990s, however, saw contraction in a far greater spread of industries and businesses, taking in the 'traditional' sectors which as coal, iron and steel, shipbuilding, **and** in sectors representative of the 'second industrial wave' of the twentieth century such as chemicals and vehicle manufacture.

Given the focus on the late 1980s and early 1990s, this study is concerned with a time period marked by a continuity in central government in Britain. Here there was not just the same political party — the Conservatives — in power, but, for a large part of the time period across which the study stretches, the same Prime Minister: Margaret Thatcher, Prime Minister from 1979-1990. Hence, this work is a study of economic regeneration in the coalfields under a series of Conservative governments. In view of the unitary governmental structure of the United Kingdom, the ultimate responsibility for policy outcomes — no matter which other organisations were involved — lies with central government. There is at least a case for arguing that the praise for successful policy

17

outcomes — heightened economic activity, stimulated entrepreneurship levels — as well as the blame for unsuccessful policy outcomes, must be laid at their door. Obviously, however, any analysis of any variety of policy has to take into account the largely unquantifiable impact of extraneous factors: in this case, the most salient would be the world recession of the late 1980s/early 1990s; or, as another example, a poor level of attainment of higher education places in localities associated with coal mining. Such extraneous factors constrain governments in what they feel able to do, but also put a constraint on what the potential outcome and impact of a regeneration policy could be.

This work also acknowledges the trend, well-established in the post-Second World War era but accelerating in the 1980s and 1990s, for deep-mined coal extraction to be concentrated primarily in the 'central' coalfields and away from the so-called 'peripheral' coalfields. The 'central' coalfields refers here mainly to the Yorkshire, north Derbyshire and Nottinghamshire coalfields, with some investments also going to West Midlands collieries such as Daw Mill in Warwickshire, or Trentham in Staffordshire, which could be classed as 'central'. The 'peripheral' were the coalfields of the north west (Lancashire, Cumbria and North Wales), South Wales, Northumberland, County Durham and Scotland. The latter were to suffer massive deindustrialization in coal in the 1960s, and further contraction in the 1980s (O'Donnell, 1988; Turner, 1985).

A comparison between the Nottinghamshire 'central' coalfield and the Durham 'peripheral' coalfield, for example, is instructive in relation to this and provides evidence of this abiding trend towards concentration in the central coalfields.

In 1947, in Nottinghamshire, there were 31,078 miners. The change by 1974 was marginal, with a reduction of less than a thousand, to 30,300 (Taylor, 1984, p.171). In the Durham coalfield, by contrast, manpower fell from 97,924 in 1958 to 44,160 in 1968 (Krieger, 1984, p.104). By 1974, the number of miners employed in the Durham coalfield had fallen to 18,420 (Taylor, 1984, p.171).

The concentration of mining in the 'central' coalfields continued in the 1980s and early 1990s. The National Coal Board's *Report and Accounts 1982/3* noted that in the financial year 1982/3, £726 million was added to the Corporation's notional value in the shape of additional fixed assets. £564 million, or 78 per cent, of this went to the 'central' coalfields, and almost half of the total additional fixed assets — 46 per cent, or £332 million — went to two of the then operational/administrative units of the 'central' coalfields: North Yorkshire and Barnsley. The former would include the giant Selby development, and the latter the rejuvenation and restructuring of a long established coalfield (National Coal Board, 1983, p.44; Barnsley Metropolitan Borough Council, 1984).

By the mid-1980s and early 1990s, most of the 'central' coalfields — as of October 1991, with the exception of north Nottinghamshire — were experiencing as drastic, if not more drastic, a contraction in the coal industry as had been witnessed in the 'peripheral' coalfields earlier. This represented the justification for the major focus within this work on economic regeneration schemes and projects in the central coalfields: most of the coal mining in Britain had been located there by the 1980s, and the scale of contraction in the 1980s and 1990s was new and swingeing.

The rationale for the choosing of particular schemes in particular localities is dealt with in more detail in the individual chapters. The primary criterion for selection was the extent of economic dependence in the recent past on the coal industry. Projects were examined where local economies had had a very high dependence on the coal industry and had suffered major and rapid contraction in that industry in the 1980s and 1990s.

What is not being claimed is that every single attempt at economic regeneration in Britain's 'central' coalfields in the late 1980s and early 1990s is examined here. That would have required an investment of time, money and labour beyond the scope of this work. The claim that is being made here is that a representative sample of regeneration initiatives has been examined, covering state involvement at more than one level of government (local, multi-local, national), and covering a variety of localities within the 'central coalfields'.

1.6 Deindustrialization

The reality of coalfield deindustrialization via the closure of coal mines, coal washeries and coal preparation plants, workshops, and other industrial activities ancillary to coal mining, has to be set within the wider context of deindustrialization which affected and afflicted parts of Britain particularly severely during the 1980s (Levie et al, 1984; Dickson and Judge, 1987), and the early 1990s.

Deindustrialization has been variously defined. For Campbell (1990), for example, it referred to the **relative** decline in Britain of the share of total economic output held by the manufacturing sector. According to his figures, growth in the service sector was nearly 30 per cent during the period 1982 to 1988, following the economic recession of the early 1980s, compared to a growth of less than 25 per cent in the manufacturing sector over the same period.

On Singh's (1977) definition, however, deindustrialization was interpreted as a progressive failure to achieve a sufficient surplus of exports over imports of manufactures.

Another interpretation was that of Bacon and Eltis (1978), who argued that during the 1960s and 1970s in Britain, what they called the 'non-marketed sector' of the economy — largely public sector services — had been allowed to grow by government authorities at the expense of the 'marketed sector.' The latter was any part of the economy wherein a business, industry or organisation had to sell its services or products to a customer in order to survive in the market economy. In the 'non-marketed sector' — health care, education, local government services, for example — the taxpayer paid for the service. These sectors were not financed by

20

selling their services in the market, rather, they were paid for out of taxes. According to the thesis, governing politicians had found it expedient to increase expenditure and thereby employment in the non-marketed sector in times when unemployment elsewhere in the economy was rising, or in danger of rising. This employment expansion in the non-marketed sector took on a 'ratchet' effect: employment rose here in times of a general economic slump, but did not decline once the slump was over. The manufacturing sector, ('producers' in the Bacon and Eltis terminology), was therefore deprived of labour power in times of increasing general economic activity. As a consequence, the domestic manufacturing sector failed to respond adequately to demand for manufactures when demand did increase. They were unable to respond adequately because of labour shortages, and this failure then led through to, either, an addition to inflationary pressures within the economy as more and more employees in the service sector chased the declining product of fewer and fewer domestic 'producers', or, the failure would exert pressure on the balance of payments as more and more imports were sucked in to meet domestic demand.

In terms of the British coal industry as represented by the NCB and British Coal, it is difficult to categorise it as belonging definitively to either the 'non-marketed' or 'marketed' sector. Although it made an operating profit between 1975 and 1981, and operating profits in the years 1986 to 1990, overall — after interest payments; monies for redundancies; depreciation; monies for repairs to subsidence damage and so on — it managed to make a financial surplus only in the three years 1976, 1977 and 1978 across the whole period 1975 to 1990 (NCB, 1981; British Coal Corporation, 1990).

Relying, therefore, for its existence on the taxpayer might lend credibility to a characterisation of the British coal industry as belonging to the 'non-marketed sector' in the 1970s, 1980s and early 1990s. On the other hand, there were always individual collieries and other operating units within the British

21

nationalised coal industry which **did** make a profit — which survived through selling their products to customers — and which would therefore be categorised, on the Bacon and Eltis dichotomy, as being part of the 'marketed sector'.

So, for example, while the then 'Western' administrative/organisational unit of the then National Coal Board made a loss per tonne of coal mined in 1981-82 of £5.50; certain individual collieries within that unit made a profit. Point of Ayr colliery in North Wales, for example, made a profit of £5.9 per tonne of coal extracted in 1981-82. Florence colliery, in Staffordshire, also part of the then 'Western' area, made a profit of £7.00 per tonne extracted. Hapton Valley, in Lancashire, another 'Western' area colliery, was profitable to the tune of £12.00 per tonne extracted in the same year (Monopolies and Mergers Commission, 1983).

In general terms, the Bacon and Eltis thesis was not watertight. It went **against** evidence presented by Cairncross (1978) which suggested a substantial **increase** in the volume of manufacturing investment — investment in the 'marketed' sector — in the 1960s. A further claim by Bacon and Eltis was that the increase in the non-marketed sector had reduced potential exports. In other words, more of the British economy's home production had gone to meet home demand. Karel Williams et al (1983) noted to the contrary, however:

> 'The British manufacturing sector is increasingly heavily committed to export business which accounted for 20 per cent of UK manufacturing sales in the mid 1950s and over 30 per cent by 1980.'

As the Bacon and Eltis thesis is not central to this work, this is not really the place to present a detailed critique. For those seeking a more in-depth analysis of the theories behind industrial contraction, the various interpretations of the concept of 'deindustrialization' were elucidated and examined with clarity by Cairncross (1979). Even

22

though deindustrialization has deepened and widened, and progressed apace across Britain since the 1970s, the Cairncross exposition retains validity.

Here, however, a central concern is deindustrialization in one industry: coal. It is that which forms the background context of a study of economic regeneration attempts. And though coal mining, as a 'heavy' traditional industry, was often related to the 'manufacturing' category — upon which most studies of deindustrialization have focused — in fact it was **not** strictly part of the manufacturing sector, but, rather, was and is an **extractive** industry. Many individual manufacturing businesses, nevertheless — large and small — have depended upon the continued existence of the coal industry for their own continued existence: mining equipment suppliers, for example, providing equipment for digging, tunnelling, transporting, winding, lighting. In fact, in 1992, the Association of British Mining Equipment Companies (ABMEC) launched a 'Campaign for Coal' (letter to *The Guardian*, 9 April 1992). Thus elements of capital, as well as sections of labour, were working at this time to save the deep-mined coal industry from virtual elimination.

Therefore the decline in mining contributed to the wider deindustrialization in the British economy. And the same has operated the other way around: a decline in heavy industry meant a decline in the quantity of energy — and, therefore, very often, coal, either directly or through the mechanism of electricity produced from coal — consumed by sections of heavy industry. The consumption of steel in the British national economy, for example, fell from 20m tonnes in 1979 to 14.6m tonnes in 1983 (Iron and Steel Statistics Bureau, 1986). Some collieries — such as the former Cortonwood colliery, the former Wath Main colliery and the former Thurcroft colliery in South Yorkshire, for example — had provided coal directly to the steel industry (Barnsley Metropolitan Borough Council, 1984).

An entire phalanx of industry was beset by closures during the late 1970s, the 1980s and early 1990s. This included the dramatic contraction in the steel industry which preceded the rundown in the coal industry; large scale cutbacks in shipbuilding, car manufacturing, textiles, and in a spectrum of engineering activities. One third of the employment in Britain in the metal manufacturing, metal-using and chemical industries were shed between 1979 and 1987, for example (Lewis and Townsend, 1989). And Levie et al (1984) noted that employment in manufacturing industry was cut by a third (about 2.4 million jobs) between 1969 and 1981, with half of the jobs lost after 1979.

The situation was so severe that some social scientists were beginning to argue that advanced western economies, and particularly Britain, were entering a post-industrial phase (see Frankel, 1987). For some this was viewed positively, and for others it had the potential to yield positive effects. Illich (1973), for example, argued that post-industrialism could and should be an economic phase which would liberate the individual:

'A post-industrial society must and can be so constructed that no one person's ability to express him or herself in work will require as a condition the enforced labour or the enforced learning or the enforced consumption of another.'

Illich's target was industrial society per se. Post-industrialism, as a move **away** from industrial society, would be a positive development so long as it was constructed in an equitable fashion, a fashion that was liberating for the individual.

Others viewed the economic change — deindustrialization — as an unmitigated 'bad': it rendered the skills people had invested sometimes a working lifetime developing, redundant; it rendered the skills — and, indeed, a whole way of life — that had been fostered in a community and passed on from generation to generation, redundant; it meant a lack of economic — and concomitant social and personal — opportunities for an aspiring, but essentially redundant, workforce (see, for example, Hudson and Sadler, 1985; Beynon, Hudson and Sadler, 1991).

Importantly, though, coal has had a life outside the bounds of manufacturing industry, and indeed, could potentially continue to have a life into a post-industrial society. For energy is needed, even in a post-industrial society, to light and heat homes, schools, hospitals, service producers of all kinds, and to power the 'new' industries such as robotics. If the energy source chosen is coal — even if that is indirectly via electricity produced from domestic coal — then a domestic coal industry could potentially survive an even deeper and more generalised deindustrialization elsewhere in the economy than was seen in the 1970s, 1980s or 1990s. That it could potentially survive did not necessarily mean that in practice it would.

1.7 Reindustrialization and the 'state'

Theoretical analysis of **why** the 'state' saw fit to embark upon certain reindustrialization measures — or, more broadly, measures ostensibly designed to promote economic regeneration — are not examined within, though presumably the 'state' could have decided **not** to take any action in relation to regeneration, had it so wished. Boulding (1989) **has** examined this aspect of the debate, in relation to reindustrialization measures in steel closure areas, where a rapid decline in that 'traditional' industry took place largely in the few years before the large scale contraction in the British coal industry in the mid and late 1980s and early 1990s. Boulding focused in particular on the Marxian theories of Claus Offe on the perceived need for states in capitalist societies to maintain legitimacy for the political system in place, and support amongst the population, whilst at the same time maintaining a position of societal and economic domination. Given that this ground has already been covered, there is little point in rehearsing it here. Nor would it be particularly germane to this thesis to offer a challenge to, or analyse in depth, the arguments advanced by Boulding. His argument may or may not have

25

validity, but it is not central to this work to analyse the veracity of the propositions pointed to by Boulding. The starting point here is that attempts at 'reindustrialization', or economic regeneration, **were** embarked upon in the 1980s and 1990s in localities where the coal industry had in the past been a dominant force in the local economy: this work accepts that as a given variable. The focus within, therefore, is on the policies themselves: factors leading to their formulation;the process of implementation; and some attempts at evaluation. And, in order to properly understand these policies and their impact, the political and economic **context** of prior deindustrialization had to be at least partially sketched.

1.8 Post-coal regeneration: factors justifying special attention

If, as asserted here, deindustrialization was a commonplace phenomena, affecting different sectors of the economy and different localities in Britain in the 1970s, 1980s and 1990s, what can be the justification here for a focus exclusively on economic regeneration attempts in localities formerly associated with the coal industry?

Firstly, although the British coal industry had been in continuous decline in terms of number of collieries and labour force levels since nationalization in 1947, albeit with a period of stabilisation and occasional slight upturn in the 1970s (Turner, 1985), contraction in the 1980s and early 1990s was particularly drastic and rapid. This resulted in a rapid change in the relative economic position of localities associated, or formerly associated, with the coal industry. Beynon, Hudson and Sadler (1991) noted, for example, that of the 15 most deprived districts in the United Kingdom in 1983, no mining district featured among them. By 1988, there were five districts associated with coal mining in the top 13.

Secondly, it is evident and obvious that certain localities traditionally associated with

26

the coal industry — often referred to as 'pit villages'— were virtually one-industry economies. The 'pit village' developed solely because the coal industry had developed. Indeed, private sector coal mine owners would organise the building of company-owned housing near the pit head in the nineteenth and early twentieth centuries (ACOM Secretariat, 1991). Company villages were created, for example, in the 1920s in Blidworth, Bilsthorpe, Clipstone, Welbeck, Edwinstowe, Ollerton and Harworth in Nottinghamshire (Field, 1986). The phenomena of the one-industry economy, though sometimes obtaining in relation to other industries such as steel or shipbuilding, was more common in relation to the coal industry. In addition, the 'pit towns' were often small and sometimes isolated. In such circumstances, where the pit closed or partially closed in a one-industry economy, the local economy would have little else to offer the local population in the way of paid employment and economic activity. This was not the case in all coalfield localities (Howell, 1989, p.6) — there were exceptions particularly in the Lancashire and Nottinghamshire coalfields — but the point held true in sufficient localities for it to be significant.

Essentially, then, it can be argued that many coalfield or former coalfield localities have had poorly developed economic structures. The Coalfield Communities Campaign (1986) argued, for example, that:

'The NCB does not appear to generate economic activity amongst suppliers to the same extent as most manufacturing industries.'

Moreover, they argued that the small business sector indigenous to coalfields was often underdeveloped:

'In the West Yorkshire coalfield area, for example, in 1981 small firms accounted for only 1 in 20 persons employed compared with 1 in 11 for the country as a whole.'

27

The one-industry nature of many localities associated with coal presented problems for any form of economic regeneration strategy. It meant that skills appropriate to other industries and businesses may not have developed or been encouraged. Though there is no readily-available evidence to support the contention, there are those who have argued that governments deliberately discouraged alternative industries from developing in the coalfields, for fear of tempting labour away from the mines. Those governments, it was argued, recognised that the extraction of coal was in the 'national interest', and wanted to ensure that there were no avoidable factors obtaining that jeopardised its production (see, for example, Coates, 1992). If appropriate skills in the labour market did not exist, inward investment from industries/businesses alternative to coal was less likely to happen, as companies would be unsure that their labour requirements would be met. Similarly, spontaneous economic regeneration following coal mine closures was less likely in these circumstances, as people were not learning skills in the established dominant industry which could then be transferred into, say, a new entrepreneurial small business which might contribute to a rejuvenation of the local economy. The skills associated with coal mining were, rather, usually specific to the coal mining industry.

Hall (1988) argued that industries which showed potential for growth in the late twentieth and early twenty-first centuries — robotics, biotechnology, 'alternative' energy systems — would be unlikely to locate themselves in the areas formerly associated with traditional industry. Instead these new industries would seek out their own environmentally-attractive locations, possibly within the vicinity of an 'entrepreneurial university such as Silicon Valley and Stanford University, or the Research Triangle of North Carolina, located between Duke University and the University of North Carolina.'

Some support was offered for this thesis by Champion and Green (1989) who studied 280 Local

Labour Market Areas (LLMAs) across Britain. LLMAs were defined by Champion and Green as:

> 'real places that are relatively independent and on which the quality of life of the local inhabitants largely depends. For present purposes this specifically-defined set of areas is much preferred over the standard official statistical areas like counties and districts, the boundaries of which pay scant attention to the functional realities of Britain's settlements.'

In the study of LLMAs, Champion and Green noted that Barnsley and Mansfield — two towns heavily associated with coal mining since well before the Second World War and into the post-War era — had a proportion of people employed in producer services and high technology industries which was 'particularly low' in relation to other LLMAs. Barnsley, for instance, had under 4 per cent of it workforce in these sectors as of 1989: a tenth of Bracknell's level. The south of England overall had 10.2 per cent of its workforce concentrated in producer services and high technology industry, though admittedly this figure was distorted by a concentration of these sectors of the economy in London (Champion and Green, 1989). **Producer Services** here can be defined as those activities such as banking, advertising, insurance, market research, scientific and professional services and research and development, which are sold to manufacturing industry or other service-orientated businesses. They can be contrasted as a category with **consumer services**, where the demand for services that is being met is demand from the household (Allen, 1988). The distinction is significant: Daniels (1988), for example, noted that service industry expansion and employment growth took place only selectively in the 1960s, 1970s and 1980s, and most of it was in the producer services sector.

Champion and Green's findings on the economic structures of LLMAs are very important. It was the 'new' economic sectors of producer services and high technology industry, that had the potential at least to revivify local economies that had been associated in the past with traditional industry. The best hope provided by the maintenance of an old industrial structure was that it would satisfactorily hold a local economy in suspended animation — an avoidance of death, but an inability to live properly — for a period, during which a broader-based local economy might be developed. More likely, perhaps, was that, in the absence of the development of a high technology industrial sector, and in the absence of the development of producer services such as the finance, insurance and banking industries, localities associated with traditional industry would see the decline of that traditional industry and its partial replacement with 'successor' old industries.

The biggest private sector industrial investments in the Barnsley locality in 1991, for example, were by Koyo Seiko and the Spring Ram Corporation. Koyo Seiko was a Japanese-owned ball bearing manufacturer supplying the car manufacturing industry. It commissioned the building of a £50 million factory in 1991. The company expected this factory to provide 250 jobs (*Barnsley Chronicle*, 2 August 1991). It was on the site of Dodworth colliery, which had closed in 1985. Spring Ram were a bathroom and kitchen furniture manufacturing group which commissioned the building of a £25 million factory in 1991. The company expected this plant to provide 300 jobs (*Barnsley Chronicle*, 26 July 1991). It was on the opposite side of the M1 motorway from Woolley colliery, which had closed in 1987. Neither of these investments were unwelcome, but neither of them could be classed as being anything other than representative of the 'old industries': they were not high technology; they were not producer services.

A possible partial exception to Hall's thesis was provided by South Wales. South Wales was home for much of the twentieth-century to the 'traditional' industries of coal and steel. In the early post Second World War era, South Wales became a location for British firms, and a few foreign firms, in the then expanding electrical engineering industry. From about 1974 onwards it became a prime location for firms, especially from Japan, producing advanced electrical products. Such inward investment continued throughout the 1980s. It was a 'partial' exception only to Hall's thesis because some of the electrical engineering sector located there would not fit into Hall's categorisation of industrial sectors where growth was seen as a probability in the future: robotics, biotechnology, 'alternative' energy systems. The advanced electronics sector, being operated in South Wales in the 1980s and 1990s by companies including Matshushita, Mitel and Siliconix, however, certainly would have. What also mattered, of course, in relation to economic regeneration activities, was the question of 'who benefited' from these activities. It has been argued that the economic recession of the late 1980s and early 1990s, and the influx of overseas firms into South Wales, led to employers there becoming more reluctant to employ workers formerly engaged in the 'traditional' coal and steel industries, except where they were obliged to in order to qualify for financial assistance from governmental authorities (Morgan and Sayer, 1988).

Another possible partial exception to Hall's thesis might be West Lothian, which encompassed Livingston new town, in Scotland. In 1968, there were four coal mines to the east, north east and south east of Livingston: Easton, Riddochhill, Whitrigg and Polkemmet (*Guide to the Coalfields,* 1968). The first three closed in the 1970s; Polkemmet survived until June 1986. Other heavy industry had also been located elsewhere in the nearby locality: the Leyland tractor plant at Bathgate was a key example. Opened in 1961, the Bathgate tractor plant was itself a product of regional policy in response to what was then seen as high unemployment in Scotland: the figure for the country for male unemployment was 5.1 per cent. The then Board of Trade persuaded the then British Motor Corporation to locate the plant — originally intended to produce tractors, lorries and diesel engines — in West Lothian rather than in its preferred location of Longbridge in the West Midlands. The government also provided a subsidy towards capital costs and arranged for the construction of the factory (*The Times*, 22 January 1960). The Bathgate plant closed in June 1986.

Despite this background in heavy industry, parts of West Lothian, particularly Livingston new town, became, from about 1984 onwards, a centre for high technology, inward investing companies from the USA, Japan and elsewhere. These included companies such as Mitsubishi, producing video cassette recorders; NEC Semiconductors, producing very large scale integrated circuits (VLSI); and Seagate Microelectronics, from California, producing linear integrated circuits for analogue and power management applications. The region suffered high unemployment in the 1980s, with unemployment in Livingston reaching 17.9 per cent in 1983. However, between 1983 and 1989, Livingston saw the location or expansion of 1523 firms in North American ownership and 1198 firms in Japanese ownership (West Lothian Business Information Bureau, 1990). Nevertheless, again, arguments were put forward that it was not the older, established working class that were being recruited by some of these firms. It was, rather, teenagers, often female, who were seen as being 'uncontaminated' by exposure to the union practices and working methods associated with the 'old' industries (see, for example, Meegan, 1988).

The difficulties outlined above provided those who sought to regenerate the coalfields and former coalfields with a series of problems. Those problems mounted if the objective of any economic regeneration strategy was to reach a 'target' constituency within the coalfields of people formerly working in traditional industry, or those who would have **expected** to work in traditional industry. Such problems would tend to militate against a spontaneous economic regeneration following coal mine closures in these localities. The existence of these problems would also tend to be dysfunctional to a regeneration strategy **stimulated** by central or local governments and/or their agents, or to public sector-private sector regeneration efforts or, indeed to efforts by other body.

Nor were these the only difficulties militating against regeneration. ACOM, the multi-local authority European-wide pressure group comprised of sub-government institutions representing coal areas noted, for example, that low levels of educational attainment were often the norm in localities associated with coal mining. This was partly because of the culture of heavy manual labour that was associated with the mining industry (ACOM, 1991).

Poor image could also be another problem militating against the chances of success of economic regeneration efforts. This might manifest itself as outsiders holding a mining locality in poor esteem, thus militating against inward investment. Or alternatively, it might manifest itself as a lack of community and individual self-confidence within the mining, or former mining, localities themselves (see, for example, Derbyshire County Council, 1991).

Poor accessibility and transport infrastructure could be another problem facing organisations seeking to regenerate coalfield areas. As of 1991, for example, the Dearne Valley in South Yorkshire was surrounded by motorways, but access to them was poor, pending road improvements in that locality (Coopers and Lybrand Deloitte and Sheffield City Polytechnic, 1990). Some of the mining communities in the South Wales valleys were relatively isolated from the east-west corridor of development along the M4 motorway bound for England (ACOM Secretariat, 1991). This situation was fostered, of course, by the fact that when large parts of the coal industry was being developed the British economy was still in the age of the canal and steam railways. When road development became more significant it coincided with the rise of newer economic sectors in different geographical areas (Parkinson, 1985).

Because something is difficult does not mean it should not be attempted. Indeed, the very fact that effecting economic regeneration in localities formerly associated with coal mining might be more difficult than effecting it elsewhere, in itself pointed to a justification for the attempt at regeneration to take place. Similarly, it pointed to a justification for the evaluation of those regeneration efforts. For, given the variety of techniques of government or quasi-government stimulated regeneration that might have been adopted, and given that virtually all of them presented a cost to the public purse, evaluation should have been at a premium. As Storey has noted:

> 'A key element of local employment initiatives today is evaluation. Any local agency needs to know which policies are cost effective in terms of achieving specified objectives and which are less effective, in order for it to allocate and reallocate its resources. Despite its importance, however, there are relatively few evaluations which can be drawn upon and their quality is not uniformly high.'

It was in recognition of this that this work was approached.

2 Policy agenda determination, the coalfields and the Coalfields Communities Campaign

2.1 Introduction: agendas and demands on governments

For a collection of issues, problems or 'demands' to have any chance of gaining the attention of governmental authorities, a recognition of those problems has to be on what can be called the political agenda. The deleterious effects of deindustrialization in the coalfields — and the arguably concomitant need for government or quasi-government initiated efforts at economic regeneration — had to achieve the status of being an item on this political agenda before there was any chance of a concerted policy response to it by central governments. By 'agenda' here is meant a list of issues and problems which the public, and sometimes the government, deems worthy of governmental attention (Benyon, 1985; Solesbury, 1976).

Cobb, Ross and Ross (1976) divided the political agenda idea into two. Firstly, there was the **public** agenda. Here, issues/demands had achieved a 'high level of public interest and visibility'. Secondly, there was the government, or formal, agenda, which consisted of 'the list of items which decision makers have formally accepted for serious consideration'. The objective of those supporting issues/demands which were prominent on the public agenda was to get recognition of them accepted on to the **formal** agenda; and, once accepted on to that formal agenda, to get policy action directed at them. Here, the term 'policy agenda' — as distinct from political agenda, public agenda or formal agenda — will be used to indicate that the issue/problem under review actually brought forth policy action.

In all modern industrial, or possibly post-industrial, societies, governmental authorities will be faced with a series of articulated policy 'demands' from different groups and individuals: for better health care provision, for example, or more resources for education, or increased subsidies for certain industries. Some demands might have limited direct financial and economic consequences for the government authorities that are being asked to accede to them. The demands for independence from colonial ties by African and Asian countries in the 1950s and 1960s might (or might not) fit that bill, for example. What is obvious in social science is that not all demands that involve the mobilisation of economic resources — usually expressed as money but referring here also to organisational inputs — can be met in the way desired by their protagonists, or indeed met at all: economics teaches us that resources are limited, so even if the will to satisfy all demands were there, demands would remain insatiable (Lipsey, 1963, p.50). Solesbury (1976), similarly, pointed out that governmental time is limited and resources usually scarce: not every issue could be attended to.

Alongside this, to satisfy the demands of some would negate the demands of others. For example, to grant publicly-funded facilities to use as abortion clinics might satisfy the demands of pro-abortion groups but would do the opposite in relation to anti-abortion groups.

If there are articulated demands, there may also be 'unarticulated' demands. Because they are unarticulated, the needs and requirements of these groups and individuals will fail to reach the political agenda, and therefore fail to merit even consideration for governmental policy action or resources. In what Bachrach and Baratz (1962) categorised as a 'non-decision', governments may decide to do nothing in relation to a particular problem or issue. This situation may arise where there is a powerful interest group — in economic or political terms or both — working to keep the demands of others off of the political agenda. An example might be a business group threatening to withdraw investment. Another way it may arise is if the individuals with demands are unorganised or weakly organised, and cannot therefore press for their demands to be met with any chance of getting them on to the agenda: the unemployed, for example, might fit this category. Some pluralists argued, however, that even the demands of the unorganised would be responded to by policy-makers, because the latter would realise that the unorganised had the **potential** to organise, and therefore the potential to exercise significant electoral pull or power in some other way (see Dearlove and Saunders, 1984). And it remained a possibility that policy-makers would respond to the demands of the unorganised through altruism. The maintenance of public support remained, after all, crucial to the continued existence of a system of government and the aspirations of a political party to government (Burch and Wood, 1990, p. 54-64). Elsewhere, radical political scientists have argued that certain demands will be kept off the political agenda by the very **structure** of the political system itself: individuals and groups will not articulate certain demands because their 'real interests' are obscured from them by the political system. The structure of the political system is such as to allow the 'real interests' of people in society to be obscured by a dominant social class in order to maintain social relations of dominance of one class over another (Lukes, 1974). As a hypothetical example, rendered simply to relate abstraction to concrete level, it may well have been in the 'real interests' of

the working class in Britain to have a Labour
government in power in the 1980s and 1990s.
Even if that was the case, and of course it may
not have been, with four consecutive general
election victories by the Conservative Party by
1992, the overwhelming majority of the working
class appeared to have been unaware of it.

To return to the pluralist perspective, it
is argued that other demands will not be met for
reasons of an absence of technical competence or
feasibility (Burch and Wood, 1990; Solesbury,
1976). As of 1992, for example, there was no
vaccine against the HIV virus: most would agree
that governments would have liked, desperately,
to be in a position where scientists had found
one; but medical technology had not advanced
that far.

The central point then is that only demands
which have been articulated by groups or
individuals who are relatively powerful, have
been accorded a priority by governmental
authorities, and where there is a technical
competence for their achievement, would be
acceded to. What serves to determine this
priority becomes, from a political science
perspective, the most important aspect. And it
is this priority-determination which forms the
political agenda; where specific governmental
action occurs, the priority-determination forms
the policy agenda.

2.2 Models of agenda building

Cobb, Ross and Ross (1976) identified three
mechanisms by which issues/problems/demands
might reach the policy agenda.

The first was the 'outside initiative'
model. Here:

'Issues arise in non-governmental groups and
are then expanded sufficiently to reach,
first, the public agenda and, finally, the
formal agenda'. (Cobb, Ross and Ross, 1976,
p.127).

A second was the 'mobilisation' model,
where issues were:

'initiated inside government and, consequently achieve formal agenda status automatically.'

In order to gain support for a policy, so as to increase the chances of success, those initiating the policy here may seek to place it on the public agenda: hence the term 'mobilisation'. An example might be a development programme in a Third World country where some kind of action or inaction (say birth control) is required from the population at large. Unless that was on the public agenda, the population at large would not know about it and therefore would not participate.

A third model of agenda building identified was the 'inside initiative' model. Here issues were placed on the policy agenda from **within** government institutions but the objective of the supporters of those issues was that the general public did not become aware of them. The issues were on the **formal** agenda, but kept off the **public** agenda. The nuclear defence policy of successive British governments might be a case in point: the decision to manufacture an atomic weapon, for example, in 1947 was taken in secret by a cabinet committee; as was the decision to explore the possibility of buying the Trident nuclear weapons system from the USA in the 1970s (Jordan and Richardson, 1987, pp.157-158).

Two factors identified by Solesbury (1976), on how issues command attention (and therefore reach the political agenda), are relevant to the subject under discussion here: 'particularity' and political ideology.

Solesbury (1976) noted that:

'Issues have particularity if they can be clearly exemplified by particular occurrences or events.'

So, in relation to environmental politics and policy, for example, Solesbury cited the following as exhibiting particularity:

'The London Smog of 1952, the demolition of the Euston Arch in 1962, the Torrey Canyon disaster in 1967 ....'

Political ideology was important in the sense that the issue/need/problem being pressed for would have a far greater chance of achieving recognition and action if it was in sympathy with the political ideology of the existing government or the more widely prevailing political ideology in society.

2.3 Agenda building and the coalfields

At least three major attempts can be identified of pressure groups seeking to place on to the policy agenda the interests, needs and demands of some coalfields residents. All of these were commenced in the 1980s. All of them would fit into the 'outside initiative' model categorisation referred to above. The most obvious and conspicuous was the 1984/85 year-long strike by the National Union of Mineworkers. That strike has been discussed extensively elsewhere (see, for example, Adeney and Lloyd, 1986; Callinicos and Simons, 1985; Coulter et al, 1984; Goodman, 1985; MacGregor, 1986), and it is not proposed that discussion of its political objectives, the way it was prosecuted, or the way it was reacted to by governmental authorities, should be continued here. Suffice to say, as an attempt at preventing deindustrialization in coal — and therefore as an attempt at placing the demand that pit closures must stop on the **formal** agenda — it obviously failed.

The 1984/85 strike had a clear precursor, however. That was the unofficial stoppage by large sections of the NUM, designed to stop impending pit closures, in 1981. From the position of the NUM, this precursor was more successful than the 1984/85 stoppage, although its modest 'success' was limited to a time-specific period: 1981 to 1984. And, indeed, there **were** pit closures during that period. The then NCB was operating 211 collieries at the end of 1980/81 financial year; by the end of the 1983/4 financial year, the figure was down to 170 (National Coal Board, 1985). Nevertheless, the 1981 stoppage can be considered to be the second major attempt to gain access on to the political agenda for the economic needs/demands of some of the residents of the coalfield areas.

February 1981 saw Derek Ezra, then Chairman of the NCB, declare that the British nationalised coal industry was extracting seven million tonnes a year surplus coal, following the general recession in the economy. The industry would have to contract: 23 pits would have to close during the 1981/82 financial year (Turner, 1985).

The response of the workforce to this closure programme was swift and erupted spontaneously, especially in the coalfields perceived to be most threatened, such as South Wales. Widespread unofficial grassroots action was followed by the threat of an all-out national strike by the NUM. The government's capitulation, when it came on 1 February 1981, was dramatic. The closure programme was withdrawn after the government had agreed to relax financial constraints on the National Coal Board. It was merely a momentary political humiliation for Mrs Thatcher's administration, however. Arthur Scargill displayed prescience at the time by arguing that talk of total victory by the miners was premature, and that government had merely postponed the date of battle (Priscott, 1981; Turner, 1985).

The real battle, of course, came later: in 1984/5. By that time the government was much more prepared for what might turn out to be a long dispute. Ian MacGregor, a more notably hard-line and less consensual manager than either of his two predecessors — Derek Ezra or Norman Sidall — had been Mrs Thatcher's personal appointee to the Chairmanship of the NCB (Goodman, 1985). Peter Walker had been installed as Secretary of State for Energy in 1983 and there are those who argue had been told to prepare for a strike in the coal industry. Coal stocks had been built up at power stations. Glyn England, who had complained about the Thatcher government's criticisms of nationalised industries (see *The Times*, 5 March 1982), was replaced by Sir Walter Marshall as chairman of the CEGB in 1982: a man far more committed to the expansion of the civil nuclear power project in Britain (a direct competitor to coal), and far closer to the political thinking of Mrs Thatcher (Goodman, 1985).

If the argument here is that the miners' strikes of 1981 and 1984/85 were attempts to place on the political agenda the economic demands of some of the coalfield dwellers, there are those who argued the converse of this. Their argument was that the 1984/85 strike in particular did not so much represent an attempt by the NUM to place the demands of coalfield communities on the political agenda — on an 'outside initiative' model — but, rather, that on the agenda of the government was the decision to crush/destroy the NUM. As a union in the vanguard of the labour movement, which had a recent history of successful industrial action, and a militant leadership, the NUM would be a natural target for a government committed to free market economic policies, and a disengagement from micro-economic, or industrial, policy, with an end to the financial subsidies that the latter often implied. After all, the government would not want any obstacles to the macro-and micro- economic policies it wished to pursue. The proponents of this thesis would argue that evidence existed to substantiate the claim that the government provoked the strike in order to achieve the objective of crushing the NUM: the strike was 'provoked' at the worst time, from the miners' point of view, for a strike — spring; the government had encouraged the CEGB to build up coal stocks; had 'prepared' the police for a confrontation, and so on (see Goodman, 1985). None of this could be proved, of course. But those who made the argument were effectively saying that the 'inside initiative' category — noted above — was operating here in agenda building. After all, the government never actually said at the time that its objective was to crush the NUM.

2.4 Coalfield Communities Campaign

As the two previously discussed attempts at forcing the demands of some of the residents of the coalfield communities on to the political agenda were essentially attempts to **prevent** economic change, and therefore belong in an economy and a world that no longer existed in the 1990s, it was the third attempt at influencing the political agenda that was more germane to this study. That attempt was a less conspicuous and more subtle attempt, which showed signs of greater longevity: the efforts of the Coalfield Communities Campaign (CCC).

The Coalfield Communities Campaign was launched in May 1985 by 54 local authorities which were in localities which had at the time present or past connections with the coal mining industry. By April 1991, the number of local authorities affiliated had grown to 91 (CCC, undated, circa April 1991) (see Appendix 3).

The Coalfield Communities Campaign was a rare species of pressure group. Though organisations such as the Association of County Councils and Association of District Councils had existed for years to negotiate on behalf of local authorities with central government, to promote the interests of certain groups of local authorities, and to give local authorities a 'voice,' local authority-based interest groups formed on the basis of past or present common association with a particular industry were few. MILAN, however, was one notable example of the genus. MILAN, started in 1985, stood for the Motor Industry Local Authority Network. It claimed to be 'an all-party consortium of local authorities and regional enterprise boards' and argued that it was 'determined to act as the voice of those [vehicle or vehicle component producing] communities, arguing for a constructive pre-planned approach to change' (*Local Economy*, 1989).

A similar example of a local authority pressure group built around an association with a particular industry would be Local Action for Clothing and Textiles, again a group established in the 1980s (Thomas, 1992).

43

The material in this section is based on a face-to-face interview with the deputy director of the Coalfield Communities Campaign in August 1991, supplemented by literature published by the CCC, newspaper reports and appropriate political science literature.

In terms of the political complexion of the affiliated local authorities, the CCC claimed all party support. It will be noted, however, that as of 1991, the CCC amounted to a pretty heavily Labour-orientated beast: only 10 of the 91 authorities were not under Labour majority control; only 2 of the 91 authorities were under Conservative Party majority control (see Appendix 3).

Nevertheless, all the three major national British political parties were represented to some degree at the local authority level. It also had support from past management and union people in the coal industry — Lord Gormley, predecessor to Arthur Scargill as President of the NUM, and Lord Ezra, referred to above — as well as the former prime minister, James Callaghan (*Financial Times*, 13 May 1985). A formal message of support was received at its launch from the Conservative former prime minister the Earl of Stockton, and it could also count amongst its sponsors the Conservative peer Viscount Caldecote; the Labour former minister of transport, secretary of state for employment and for social security, Barbara Castle; and Baroness Burton of Coventry, a former chairperson of the Domestic Coal Consumers Council (*The Guardian*, 15 May 1985).

Unlike the two previous 'energy based' attempts — ie, by pursuing the objective of preventing pit closures — to place demands on the political agenda, the CCC had much wider objectives, encompassing the retention and development of the coal industry, but also emphasising the need for broader economic development in the post-coal industry phase. It stated its objectives as being to:

'Help protect the jobs that remain in the coal industry'

and, importantly, to

44

'Promote the economic, social and environmental renewal of coal areas' (CCC, undated, circa April 1991).

The CCC concerned itself primarily with researching into the economic and social problems and needs of the coalfield areas — which could, in Hogwood and Gunn's (1984, p.73) terms, be seen as a voluntary association carrying out a formal 'issue search' for items to place on the political agenda — and lobbying.

The rationale behind lobbying is obvious. Unless a group gets heard — unless a demand is articulated — governmental authorities will either not know of the need to respond, or feel that the pressures upon them to respond are so weak that the problem is insignificant (see Grant, 1988).

Thus research and lobbying went hand in hand at the CCC. The findings of the former were used as material to deploy in the latter. For example, the very often negative experiences of former coal miners in Barnsley and Wakefield seeking employment after pit closures were researched into in the late 1980s, and reported on, supporting the CCC's lobbying efforts directed towards increasing investment levels coming in to coalfield areas (Witt, 1990). Similarly, the allegedly adverse impact of the potential privatisation of British Coal on mining localities was researched into (Fothergill and Witt, 1990), as were the allegedly adverse macroeconomic implications for the United Kingdom economy of increasing coal imports (O'Shaughnessy, 1990). Other reports have included a largely positive assessment of the qualities of former mineworkers in relation to their abilities in new jobs outside the mining industry (PA Cambridge Economic Consultants, 1990), billed by the CCC as 'an essential tool for those promoting new investment' (CCC, undated, circa 1990); an assessment of the contribution of coal to the 'greenhouse effect', which argued that a policy of reducing coal burn as an attempt to reduce the 'greenhouse effect' would be misguided, and argued instead in favour of greater energy efficiency (O'Keefe et al, 1989); and an attempt to investigate some of the land-use problems — potential development land 'trapped' behind greenbelt planning regulations, land being held in the ownership of British Coal long after it had ceased to be used by British Coal — facing those attempting to promote economic development in coalfield areas (Roberts and Green, 1990), amongst other work.

The lobbying by the CCC was aimed at any person or organisation that might have the potential to be influential. At the national parliamentary level, with the assumption that Labour MPs were on their side anyway, that might be Conservative MPs; at the European level, it would be MEPs or officials of the Commission of the European Communities.

The attempt at reaching the political agenda by the CCC can be compared to, and contrasted with, the earlier attempts to articulate the demands of some coalfield inhabitants by the mechanism of NUM strikes discussed above.

In political science terms, Hall et al (1975) argued that governmental authorities employ three main criteria to determine their priorities for policy action (and hence the policy agenda, as the term is employed here): legitimacy, feasibility and support.

2.5 Legitimacy

Legitimacy, here, is related to whether or not, in a normative sense, a government should be involved with a particular issue, and whether or not the demands being articulated are considered 'legitimate' from the point of view of government.

The condemnation by Prime Minister Thatcher of the leadership of the miners during the industrial dispute of 1984/85 as the 'enemy within' in July 1984 (Samuel et al, 1986; Scammell, 1986) was an indication in the strongest terms that the Thatcher government did not see that particular attempt to articulate the interests of some coalfield dwellers as even remotely approaching the 'legitimate.' Mrs Thatcher had drawn a parallel in the July 1984 speech between the battle to regain the Falkland

Islands from occupation by Argentinian forces in 1982, and the battle for the 'management's right to manage' which she believed was threatened by the 1984/85 miners strike: the message was that Arthur Scargill, the NUM's President, had the same extent of 'legitimacy' as Leopold Galtieri, the leader of the Argentinian military junta, had in 1982. Not that there had been any secret before or since the start of the strike about which side — management or union — the Thatcher government's sympathy was with.

The all-political party nature, the inclusion of the Nottinghamshire, South Derbyshire and other midlands local authorities — where the majority of miners had worked during the 1984/85 dispute — within their organisation and the inclusion of 'respected' former coal industry figures such as Gormley and Ezra, were clearly part of an attempt by the CCC to appear more 'legitimate' in the eyes of the Conservative government in power in the late 1980s and early 1990s.

The more controversial, but more contemporary to the period, figures of Scargill and MacGregor, for example, were not part of this organisation. Symbolically, in that sense, the CCC could be viewed as a yearning expressed at the level of local politics for the return of a more consensual era in industrial politics.

Indeed, *The Guardian* newspaper took the launch of the CCC in 1985 as an opportunity to criticise the NUM and its leadership. In the leader column in May 1985, the NUM was criticised for keeping a distance from the CCC and, with possibly less than a sound basis, for:

> 'regard[ing] plans for alternative job generation as part of a sell-out because such plans imply the need for some pit closures.' (*The Guardian*, 16 May 1992).

It could be argued that the criticism was wide of the mark. The CCC's response to *The Guardian's* criticism of it and the NUM for not having any NUM representatives was that the NUM were not asked. The reason for this, it was stated, was not that the NUM would have damaged the 'legitimacy' of the CCC but simply that they were a local authority organisation, **not** an energy-based pressure group organisation, which the NUM were. Elsewhere, however, it was argued that the CCC deliberately kept an arms-length relationship between itself and the NUM because of the 'intense antipathy that existed between the Conservative government' and the union (Gladstone, Geddes and Benington, 1992).

The CCC could hardly have been less legitimate in the eyes of the Thatcher government than the NUM, particularly during or since the Scargill-led strike of 1984/85. Here was an industry which represented in microcosm all that the Thatcher government believed to be the conditions that had held back economic progress in Britain: it was a monopoly in public sector ownership; its workforce was one hundred per cent unionised; mineworkers were prone to local and national strikes which were often prosecuted successfully; it was regularly in receipt of cash subsidies from government to bail it out (Turner, 1989).

If the CCC had more legitimacy than the NUM, they did not appear to have achieved full legitimacy in the eyes of central government. As an indication of this, the Autumn 1990 newsletter of the CCC recalled how the UK government had described its position on the CCC's attempts in the mid- and later 1980s to persuade the Commission of the European Communites to assist mining areas as being 'positively neutral.' In a later letter from Douglas Hogg MP, then a minister at the Department of Trade and Industry, to Hedley Salt,then National Chairman of the CCC, after the agreement in 1989 by the EC to provide extra assistance to mining areas in the form of RECHAR, Hogg commented:

'the government's attitude was neutral, though positively so' (CCC News, 1990).

2.6 Feasibility

'Feasibility' was another criterion identified by Hall et al as having a bearing on the determination of the policy agenda. They isolated two aspects to feasibility. The first related to the actual 'state of the art': the distribution of theoretical and technical knowledge. In other words, if a policy objective was to be realised, it must be within the realms of the possible. A second, and important aspect, related to the ideological and value positions of those deciding which policies to pursue, how many resources to distribute to competing claims on government, and when to pursue particular policies. In relation to such factors, policy makers will always face a range of possibilities. As Hall et al phrased it:

> 'Particular ideologies, interests, pre-judices and information will affect the kinds of conclusions which are drawn about the feasibility of different alternatives.' (Hall et al, 1975 p.479).

This, for example, along with the limited amount of resources made available to re-generation agencies in given territorial areas (dealt with in the main body of the text) (see also, Turner, 1992a; Turner, 1992b; Turner 1992c), almost dictated that a major strand of the regeneration effort in the late 1980s and early 1990s would be the attempt to revivify the small business sector and encourage entrepreneurship. For it is not that the latter in any circumstances could have been argued to be the one and only inevitable policy response: there were alternative options for regeneration, even if these were ideologically unacceptable to the central government in power at the time. Two potential responses which fitted the category of alternative policy action would have been to continue subsidies to coal mines — see the section on the enterprise zone in South Kirkby later — or to initiate a large scale programme of public works. Both, technically, would have been feasible on the first part of the definition above; though both would have been ideologically unacceptable to the Conservative governments of the 1980s and early 1990s.

2.7 Support

The other criterion advanced by Hall et al (1975) as being a necessary condition before policy action would be implemented was 'support'. Support here referred to the electoral and political consequences for governmental authorities of acting — or not acting, as the case may be — in a particular way. To extend the concept, all governmental authorities in a polity such as Britain will seek to maintain general support for the political system, as a threat to it — via civil unrest, civil disobedience or revolution — could pose a threat to their own continued existence and political reproduction, and to the economic and social stability of the country. As Hall et al (1975) had it:

'Perhaps we should refer to the net implications for support of a government promoting or ignoring an issue. We should also consider the comparatively short time perspective within which such calculations seem to be made. But in whichever way reference to the support criteria is qualified we contend that it constitutes a permanent and initial hurdle for all issues [to surmount before being lodged on the policy agenda].'

What is important, of course, from the point of view of governmental authorities is to retain or win the support of groups and autonomous individuals that have a key bearing on the maintenance and reproduction of the governing authorities' political and electoral success. These groups and collections of individuals may be identified from amongst the following. Firstly, there are those who may have a bearing on the determination of levels of investment, and consequently on economic growth and employment levels, in the economy. In all economies which retained an element of the private enterprise system — which meant, in the early 1990s, almost all the countries in the world — this political objective of governmental authorities would remain important, and in the majority of such economies the target groups and individuals could include within their representation both domestic and foreign capital. Secondly, a political party, within or outside government, needs to maintain the support of those individuals and groups which actually fund the political party (in political systems where this function is not carried out by the state). In the late 1960s, it could be argued that it was at least partially the perceived threat in this direction (together with ideological hostility from many Labour MPs) which led to the abandonment of the proposals outlined in Labour's *In Place of Strife* White Paper, even before it could be put to the House of Commons as prospective legislation (MacDonald, 1976). The Conservative Party in Britain has had, of course, to pay heed to the need to maintain support amongst its own funding constituency, largely in the business sector. Thirdly, in an elective democracy — though even less directly perhaps, within a non-democratic system — governmental authorities have to maintain or foster a sufficient general level of support amongst the electorate and citizens to continue to survive and be returned to national office.

There are some examples of governmental authorities pursuing activities and policies despite the fact that they might not engender support in general society. Hall et al (1975, p.486) cited in Britain the abolition of capital punishment and the reform of laws on homosexuality in the 1960s, and the introduction of health service charges, as three examples of governmental actions which have had mixed consequences for support, and noted that governments weigh the gains and losses to their support levels from such activities. The implication was that ideological beliefs would sometimes inform and inspire policy action even if there was a threat to sections of support.

It has to be argued, however, that the general rule is likely to be that where a central government has no support to lose by not pursing a particular policy, then political forces will militate against the pursuance of that policy. In many coalfield, or former coalfield areas, a case could be made that this 'law' of political science seemed to be in application.

It was certainly the case that the Conservative government had little to lose politically in many coalfield areas, and therefore a fairly low political incentive to accord regeneration a policy priority. In such circumstances it becomes comprehensible, if not condonable, that the Conservatives — or anybody else — might accord a priority to groups of people within which there was at least a chance that political support might be won. In some coalfield localities, such as Barnsley, for example, the Conservatives had no presence at the national parliamentary level (out of three parliamentary seats which, from 1983 onwards, incorporated the title Barnsley) (Turner, 1988). Hemsworth, near Wakefield, another parliamentary constituency heavily associated with coal mining, frequently recorded the largest Labour majority in general elections, from 1950 to 1974 remaining over 30,000 (Waller, 1983). Though there is no quantified evidence, it is generally accepted that a majority of miners in the inter-war years, and in the post-Second World War era, have traditionally been loyal to the Labour Party. For example, Butler and Stokes (1969), as part of a wider election study published in 1969, carried out a survey of parliamentary constituencies with a heavy input of coal mining: 'mining seats.' The finding was that Labour was strong in these areas even amongst those who were not manual mineworkers:

'The Labour Party received fully 50 per cent of the middle class vote and lost only a fifth of the vote of the working class.'

Loyalty to the union — the Miners' Federation of Great Britain, or its successor, the National Union of Mineworkers — was a characteristic of miners (and provided the basis of its strength, where that term could be accurately applied in the 1970s) (Allen, 1981, p.62). The union, in either of its manifestations, was loyal to Labour from about the 1930s onwards (Taylor, 1984, pp.103-132).

If the general rule is that governments will not act in particular ways where, by not acting in a certain way, they have nothing to lose, the converse is also true. They **will** act in certain ways if that has the potential to generate support amongst groups or collections of individuals which are important to that governing authority on the criteria outlined above. So, it is clear that a regeneration ethos in the coalfields which had as its objective the 'strengthening' of the capitalist ethic, by focusing on revivifying the small business sector and encouraging entrepreneurship was likely to, in itself, find support and be identified as being of key importance to the Conservative Party. It could be argued that this ideological motivation informed policy in relation to the regeneration of the coalfields over the same time period.

Thus, for example, the attempt at the 'rejuvenation' of the Doncaster Chamber of Commerce and Industry, which formed a part of the regeneration efforts of the civil servant-staffed Doncaster Task Force which existed between 1987 and 1990 (see later in text) (Turner, 1992, b), would fit into this category of revivifying small businesses. Similarly, the provision by central government of general industrial subsidies to private sector companies via the policy mechanism of an enterprise zone at South Kirkby in West Yorkshire, where benefits were administered to companies between 1981 and 1991 (see later in text) (Turner, 1991) would also fit into this category. Another example would be Conservative government establishment and support for British Coal Enterprise, which began life in 1984 as the job creation arm of British Coal (see later in the text), which took as one of its major regeneration strategies the fostering and encouragement of self-employment and small businesses (Turner, 1992c). The transformation of someone into an 'owner of a business', where it was successful, no matter how small the business, might at least militate against support being given to the Labour Party, the Conservative Party's main political rivals at the time of writing in the post-Second World War

53

era, even if it did not directly militate in favour of increasing support for the Conservatives themselves.

Another means identified by Hall et al (1975) of generating support for policy action to be directed at an issue was the hitching of one issue — by the protagonists for action on that issue — to another issue, where the latter is regarded as being important by those taking decisions and wielding political power. Wittingly or unwittingly by the local authorities concerned, it can be identified that this issue-joining strategy was employed with some success by the local authorities of Barnsley, Rotherham and Doncaster in 1991.

Michael Heseltine, Secretary of State for the Environment in 1991, had been identified since his earlier incarnation as Secretary of State for the Environment between 1979 and early 1983, as according a priority to inner urban regeneration (Lawless, 1989, p.66). Indeed, a few weeks after the first Thatcher election victory in 1979, Heseltine was making it publicly clear that fighting inner city deprivation would be a priority for the Conservatives (*Daily Telegraph*, 14 June 1979). His appointment, unofficially, as 'Minister for Merseyside' following the 1981 riots in Toxteth, also bolstered this image of being a politician keen to bring about urban regeneration. And apparently, there was more than 'image' to it: those who were involved with him in some way during the 'Minister for Merseyside' phase believed he had a sincere commitment to the cause (Parkinson and Duffy, 1984). As part of this commitment, Heseltine promoted a new innovation in urban policy in early 1991. Under the 'City Challenge' scheme, 15 local authorities were invited to compete against each other for a share of £375 million of central government-allocated money earmarked for urban regeneration over a five year period. One of the 11 'winners' from the 15, and because of its success designated a 'Pacemaker' authority by central government, as all winners were, was the Dearne Valley Partnership. The Dearne Valley Partnership was created in the early 1990s, following a recommendation by economic

54

regeneration consultants (see Coopers and Lybrand Deloitte and Sheffield City Polytechnic, 1990), as a means of co-ordinating regeneration efforts in a locality — the Dearne Valley — which was under the separate administrative jurisdiction of three local authorities: Barnsley, Doncaster and Rotherham. The Dearne Valley Partnership was comprised of Barnsley, Doncaster and Rotherham local authorities; local business people; and representatives of the Department of Environment, Trade and Industry and Employment (*Barnsley Chronicle*, 20 September 1991).

Despite winning the 'City Challenge', the Dearne Valley is not a city. Indeed, it was described by the consultants referred to above as:

'A naturally beautiful lowland river dale. Even now, most of its area is neither despoiled nor built on.' (Coopers and Lybrand Deloitte and Sheffield City Polytechnic, 1990, p.410).

What was usually referred to as the Dearne Valley comprised the small towns and villages of Thurnscoe, Goldthorpe, Bolton-on-Dearne, Mexborough, Conisborough, Swinton, Wath and Brampton, which straddled the boundaries of Barnsley, Doncaster and Rotherham. What they had in common was one time extensive coal mining and the River Dearne.

Despite this social, environmental and geographical dissimilarity from the status of 'city', by attaching to, and making the connection between, localities suffering a decline in coal mining activities and inner cities, the Dearne Valley Partnership was able to win £37.5 million over the five years from 1992 from City Challenge and a further £100 million from other government sources (*Barnsley Chronicle*, 2 August 1991).

It is instructive to examine the apparent reasons for the Dearne Valley Partnership's success in attracting money from central government.

Certainly, the Dearne Valley had, in the 1980s and early 1990s, economic, physical and

social deficiencies which in themselves might
initiate, provoke or justify governmental
activity aimed at regeneration in the economic,
physical and social dimensions. Coopers and
Lybrand Deloitte and Sheffield City Polytechnic
(1990), for example, noted in July 1990, that:

> 'In 1976, coal mining employed over 11,000
> workers in the Dearne. The figure today is
> 550, out of total employment of 15,600.
> Employment is concentrated in construction
> and manufacturing where national employment
> is unlikely to grow quickly. The fast
> growing private service industries are small
> in size;

> 'We have been told in our consultations with
> local organisations that many workers in the
> Dearne are demoralised by long term un-
> employment, that young people have never had
> jobs at all, that a whole generation is a
> lost cause, or that a profound change in
> attitudes is needed to get people back to
> work;

> 'There are widespread skill shortages in the
> Dearne;

> 'Unemployment in the Dearne is well over
> twice the national average and there is
> considerable concealed unemployment,
> especially of women who are no longer
> eligible for benefit.'

The existence of these economic and social
indicators would have provided sufficient
justification, as noted above, for governmental
authorities of almost any political hue to
intervene to attempt to stimulate regeneration.
In political science terms, however, the success
of the issue-joining strategy by Barnsley,
Doncaster and Rotherham local authorities, can
be conceptualised as meeting the following
specific outlined by Hall et al (1975, p.491):

> 'It can modify the legitimacy of one issue
> by linking it with or divorcing it from
> issues with different levels of legitimacy.'

The objective of inter-city regeneration was legitimate to Heseltine; linking the issue of regenerating small, declining coal mining towns to that increased the legitimacy of the Dearne Valley Partnership's bid.

There was another aspect to this 'legitimacy' criterion. The Conservative government which came to power in 1979 placed an emphasis on the relative efficiency of the private enterprise sector vis-a-vis the public enterprise sector (see Turner, 1989). This belief that the private sector was more efficient than the public sector ranged across a host of policy areas: ownership of industry; provision of housing; execution of 'public' services at the local level; economic regeneration; to name just a few.

Consequently, a factor which heightened the legitimacy of the Dearne Valley Partnership's bid for City Challenge funds was, apparently, the involvement of the local private sector. According to Roger Watkinson, President of Barnsley Chamber of Commerce in 1991:

'Michael Heseltine contacted the Chamber some time ago to ascertain the involvement and support of the private sector. We were able to inform him that the Chamber was fully behind the bid and was working in close partnership with our local authority.

'A major reason behind the bids [sic] success is the support of the private sector and the Chamber, as representatives of that sector, will continue to work closely on the project' (quoted in *Barnsley Chronicle*, 2 August 1991).

The acceptance by central government in 1991 of the legitimacy of the Dearne Valley Partnership's bid for funds under the City Challenge scheme reflected at least a partial acceptance on to the government's policy agenda of the need to direct policy at regenerating localities previously dependent upon coal for their economic livelihood. The establishment of British Coal Enterprise by central government, an enterprise zone at South Kirkby in West

Yorkshire in 1981, and the establishment of the Valley Regeneration Towns Project in 1988, were also examples of an at least partial acceptance of the legitimacy of regeneration issues here.

At least two aspects of policy outcome and politics relating to this issue, however, reinforced the argument that the acceptance on to the government's policy agenda of the need to respond to the plight of coalfield and former coalfield localities was only **partial**, as opposed to full.

Firstly, there was the dispute which flared between British central government and the Commission of the European Communities between 1989 and 1992. This dispute was over funds totalling around £109m over three years (*Daily Telegraph*, 26 July 1991) allocated by the European Community to declining coal mining areas, under the RECHAR initiative.

2.8 Additionality

The dispute centred on the concept of 'additionality.' The dispute looked like being settled in early 1992 when the British government indicated it would backtrack on its negotiating position. By September 1992, however, the CEC and the British government were still in dispute, with the latter party having failed to pay out a sum of £38m it had received from the RECHAR fund (*The Guardian*, 26 September 1992).

'Additionality' was fundamental to the granting of all 'Structural Funds' to member states, although the Commission openly stated that 'the additionality principle is not easy to define in a precise way' (Commission of the European Communities, undated, circa 1989).

Officially, however, to satisfy the additionality criteria, funds from the European Community should have 'a genuine additional economic impact in the regions concerned and result in at least an equivalent increase in the total volume of official or similar (Community and national) structural aid in the Member State concerned, taking into account the macro-economic circumstances in which the funding takes place' (Commission of the European Communities, undated, circa 1989).

In plain language, then, to satisfy the criteria of 'additionality', firstly, funds spent would have to be additional to what was going to be spent anyway by national or local governmental authorities; and secondly, the volume of European Community funds would have to be matched from another, governmental source: other monies from local governments or central governments, or in some cases, from public sector-owned companies.

The last part of the quotation above rendered a vagueness, however, as to what constituted an achievement of satisfying 'additionality': presumably, if 'macro-economic circumstances' in a member state were sufficiently bad, the CEC might waive a strict interpretation of 'additionality.'

In any case, later in the EC Commission document, it was made clear that the Community had established what it called a 'basic rule' of additionality, 'according to which the minimum requirement is that national spending remains constant in real terms, so that at least the EC's increased spending is additional, relative to the base year' (Commission of the European Communities, undated, circa 1989).

In terms of RECHAR money for the coalfields and former coalfields of Britain, the dispute between the British government and the CEC centred on two connected issues. Firstly, apparently the British government argued that an allowance for what would come to local authorities in Britain by way of European Community funds had been built into the determination of the level of grant from central government to local government anyway. And secondly, the Commission appeared not to believe that money from the RECHAR source would actually reach the coalfield areas. The inference was that it would simply go to relieve other pressures on the British Treasury. Bruce Millan, EC Commissioner for regional policy during the relevant period, indicated in July 1991 that this was the case:

'I'm not satisfied that the arrangement in the UK ensures that the funds have the impact in the areas that they are intended' (*Daily Telegraph*, 26 July 1991).

In December 1991, Millan took the same line:

'I am not in the business of stopping or blocking money to the UK, but I am not allowing the rules on regional aid to be breached. I have warned the British government of this for several months now. I am not willing to release money until they stick to the rules they agreed in 1988' (*The Guardian*, 18 December 1991).

This excursion into the dispute on additionality between the British Conservative central government and the European Commission sheds light on the British government's policy priorities. It seemed to indicate that, for a number of years in the late 1980s and early 1990s, a greater priority was attached to the control of public expenditure — British local authorities could not, within tight central government guidelines on their spending, 'match' funds from the European Community — rather than the 'needs' of declining coal communities.

The second piece of 'evidence' to suggest that the 'needs/demands/wants' of some of the inhabitants of Britain's coalfield and former coalfield localities found only a partial acceptance on the Conservative central government's policy agenda was the modesty of the impact of job creation schemes: in the main body of this text, this is dealt with more extensively (see, also, Turner, 1991; Turner, 1992a; Turner, 1992b; Turner, 1992c).

This 'modesty' of the impact of job creation schemes had to be set aside the depth of contraction of the coal industry in the late 1980s and early 1990s, in order to gain a true appreciation of the extent of 'partiality' of the acceptance of coalfield 'needs/demands/ wants' on the government's policy agenda. The Coalfield Communities Campaign failed to achieve recognition for coalfield demands on the government's policy agenda — like the National

Union of Mineworkers — that there should be a halt to the continued run down in the coal industry. Pit closures continued throughout the late 1980s/early 1990s on a rapid scale. Within the space of ten days in September 1991, for example, British Coal announced the closure of Murton colliery in County Durham with the loss of 900 jobs (*The Guardian*, 12 September 1991); the closure of Cresswell colliery in north east Derbyshire which had employed 565 in late 1991 (*The Guardian*, 5 September 1991); and gave the 1700 workers at the Keresely colliery near Coventry six weeks to increase productivity or face closure (*The Guardian*, 2 September 1991). Keresely closed in October 1991 (*The Guardian*, 17 October 1991).

The same story continued in the following months. In November 1991, Askern colliery, near Doncaster, closed shedding 460 jobs (*Doncaster Star*, 23 November 1991). The same month saw the announcement of the closure of Thurcroft colliery, near Rotherham, which had employed at that date 660 (*The Guardian*, 29 November 1991). In May 1992, 152 miners attempted to buy the pit from British Coal (*The Guardian*, 23 May 1992). By July, British Coal had backed out of the Thurcroft deal, stating that the miners had failed to pay maintenance costs (for an account, see Gladstone and Turner, 1992).

These closures were swiftly followed by the closures in February 1992 of Bickershaw colliery in Lancashire with the loss of 620 jobs; Sherwood colliery in Nottinghamshire, with the loss of 800 jobs; and the closure of Allerton Bywater colliery in Yorkshire, with the loss of 700 jobs (*The Independent*, 1 February 1992). March 1992 saw the announcement of the closure of Cwmgwili colliery in South Wales (*The Independent*, 28 March 1992); May 1992 the announcement of Markham Main's impending closure, in the Yorkshire coalfield (*Yorkshire Post*, 6 May 1992); June 1992 the announcement of the closure of Vane Tempest colliery in County Durham (*The Independent*, 13 June 1992). The list went on and on.

A third piece of evidence to support the argument that the 'needs/demands/wants' of the coalfield communities had achieved, at best, a partial acceptance only on the policy agenda of British central government emerged in September 1992 from within the ranks of the government itself. The evidence was in a letter written on 1 September 1992 from Industry Minister, Tim Sainsbury, to the Chief Secretary to the Treasury, Michael Portillo. In the correspondence, Sainsbury made clear his view, according to the letter views shared with Tim Eggar, Department of Trade and Industry minister responsible for coal at the time, that the government's planned response to the problems that would be caused by further pit closures was inadequate. The letter was leaked to the NUM, and detailed plans for the closure of 30 pits: over two thirds of the remaining industry. It was subsequently published in *The Guardian*, (*The Guardian*, 18 September 1992). It corroborated the point that central government was giving a higher policy priority to the control of public expenditure than it was to alleviating problems in coal closure areas. Importantly, the letter also argued that the government's response to the problems caused by pit closures compared 'very unfavourably' to its response to decline in steel making and shipbuilding areas. It noted, for instance, that after the closure of Ravenscraig steel works in Scotland, over a £100m of government assistance had been made available to that locality, and this was additional to assistance that it was eligible to anyway under regional policy by its already existing designation as an assisted area. Some coalfield areas, such as Derbyshire and Nottinghamshire, did not have assisted area status in September 1992. The letter also noted that assistance to the coalfield areas had been, and was planned to be, less than the assistance that had gone to declining shipbuilding areas of Barrow and Whitehaven and Sunderland. Sunderland, the letter noted, had received a £45 million special programme. Barrow and Whitehaven, Sainsbury noted, was a non-assisted area (under regional policy) with unemployment below the national average, but had nevertheless received a special English Estates £15 million programme. English Estates was a government agency operating on behalf of the Department of

Trade and Industry which developed industrial and business properties and sites. The letter noted that, in response to an earlier letter from Eggar and Sainsbury, Portillo was:

'not able to agree to our proposals that we should approach the Commission for further RECHAR money: nor that we should announce that English Estates would be exploring a possible programme of property/site provision in closure areas currently non-assisted.'

2.9 Power and pressure groups

This 'failure' of the Coalfield Communities Campaign to halt coal mine closures was not surprising, for, although the CCC represented a combined, or multi-, local authority pressure group, the ability of local authorities to influence national policy in almost any direction was severely limited in the mid- and late 1980s and early 1990s (see, for example, Chandler 1991). Put bluntly, the CCC had virtually no influence over energy policy nationally, and it was decisions at that level which would influence whether or not a domestic coal industry would survive in Britain. Indeed, even the influence of central government had diminished here following the privatisation of the electricity generation and supply companies in 1990. Individual local authority members of the CCC may have been able to boost coal consumption and therefore demand for it marginally by heating municipal buildings using coal, but even that limited power was weakening in the late 1980s and early 1990s with the introduction of compulsory competitive tendering by the Conservative government for local authority work. This had the effect of rendering decision-making a more and more decentralised activity, more and more in the hands of private companies as the management of leisure centres and other local authority buildings passed to the private sector (Butler et al, p.67-8, p.102). Unlike some local authorities, the private sector would not even consider burning coal unless it was the most cost effective option.

The CCC had no influence over which fuel was used to produce electricity, or indeed **where** (which country) that fuel came from. A report prepared in the early 1990s by N M Rothschild, merchant bankers, which remained secret as of August 1992 to those outside the government and the highest levels of the coal industry, apparently suggested that only about 12 British Coal mines might survive the privatisation of the industry (*The Guardian*, 3 March 1992). The legislation to allow for this privatisation was passing through parliament in 1992. As was noted earlier, opencast coal mining had also been expanding in Britain in the 1980s and early 1990s, and by 1992 it began to look increasingly likely that much of the future coal needs of the British economy would be met from imported coal and domestic opencast coal.

Surveying the coalfields in the late 1980s/early 1990s, the only defensible conclusion that the political scientist could reach was that the 'needs/demands/wants' of the coalfield areas, with one possible exception, did not figure prominently on the central government's policy agenda; the most that can be said is that they managed a partial acceptance. The possible exception was South Wales. The coalfield of South Wales was the only coalfield which saw a concerned, wide-ranging, central government-inspired, co-ordinated and sometimes implemented regeneration strategy (Romaya and Alden, 1987 and 1988; Welsh Office, undated; Welsh Office Information Division, 1988). To say such is not to argue, necessarily, that it was effective. That is a different debate. The coalfield areas other than South Wales saw a series of, relatively low-key in terms of resources and political salience, ad hoc responses sometimes initiated by a central government-established but managerially independent body such as British Coal Enterprise (Owen, 1988; Turner, 1992c); sometimes initiated by local authorities acting alone, or in conjunction with private sector partners such as Barnsley Metropolitan Borough Council's 'partnership' with the construction group Costain (*Barnsley Chronicle*, 4 October 1991).

Another example of a local authority-initiated regeneration scheme was provided by Bolsover District Council's decision to launch and partly finance Bolsover Enterprise Agency (Turner, 1992a). Sometimes regeneration schemes were central government-initiated territorially-specific policies such as the enterprise zone at South Kirkby in West Yorkshire (Turner, 1991), or the Task Force in Doncaster (Turner, 1992 b). All these projects are examined more extensively later in the text.

If the 'needs' and 'demands' of the coalfield communities failed to make more than a partial impact on the policy agenda nationally, at European Community level they met with more success. Indeed, the CCC claimed a major share of the credit for winning the RECHAR money and regeneration initiative from the European Community (CCC News, 1990). It was not definitively clear that the CCC were responsible for this 'success': others cited as a major determining factor in the granting of RECHAR the appointment of Bruce Millan as a member of the Commission of the European Community in 1989. Millan had been Labour MP for Glasgow Craigton between 1959 and 1983 and Glasgow Govan 1983-88. He was known to be sympathetic to the problems faced by communities in localities of declining coal mining.

Another example of the plight of such localities reaching the political and possibly, as defined here, the policy agenda of the Commission of the European Community was the agreement of the EC Energy Commissioner, Antonio Cardosa E Cunha, to meet a delegation from the British coal trade unions, the Coalfield Communities Campaign, and the International Miners' Federation in July 1992. The purpose of the meeting was to persuade the Commission to open up European Community markets to British coal. It was thought that such a move might save twenty British Coal mines from an otherwise swift closure (*Yorkshire Post*, 13 July 1992). The very fact that the Commissioner was prepared to meet the delegation is indicative of the fact that either or both of two issues had reached the agenda at EC level: the plight of coalfield communities; the responsible husbanding of the energy resources of the Community.

One other aspect of the CCC's activities is worth some consideration: they took a conscious decision in the 1980s and onwards to concentrate lobbying activity at the European level. Partly this was a rational response to the continuation in governmental office in Britain of a Conservative government whose policy priority appeared to be the reduction in size of the coal industry to a point where it reached 'market clearing levels' (Turner 1985), over and above the economic and social 'needs' of the coalfield communities. In blunt language, the CCC thought, probably rightly, that they had more chance of success for their campaign at European level. Partly, also, the CCC's targeting of the agenda at European level reflected the fact that through the 1980s and early 1990s, decisions on what used to be called regional policy were moving away from being made at the level of the nation state and towards being made at the level of the European Community (see, for example, Makay, 1992).

Judging the efficiency of a lobbying group such as the CCC is not easy. Firstly, this is so because one cannot know what would have happened in its absence. Perhaps exactly the same policies, or lack of policies, would have emanated from British central government, or from the European Community. Secondly, as the CCC readily confirmed, the CCC could not do anything on behalf of any **individual** locality: for to do that would be seen to be acting discriminatingly within its own ranks. The corollary of this was that it was **individual** local authorities that developed relationships with British central government through which they lobbied for 'positive discrimination' assistance: an example would be Barnsley Metropolitan Borough Council's successful City Challenge bid in 1992 (separate from the successful Dearne Valley partnership bid mentioned earlier) (*Barnsley Chronicle*, 17 July 1992). Or another example might be the designation of the formerly heavily-mined Hemsworth/South Kirkby area (under the administrative jurisdiction of the City of Wakefield Metropolitan Borough Council) as an 'intermediate area' for the purpose of national regional policy until the early 1980s (see later in the text).

Through its conferences, research publications, representation through the broadcast and print media, and its own pressure group identity, the CCC served with some success to give the coalfields and former coalfields a 'voice'. In this context, Hirschman's (1970) definition of 'voice' is useful, even if 'government' and 'government policy' should be substituted where he used the term 'management':

> 'Voice is here defined as any attempt at all to change, rather than to escape from, an objectionable state of affairs, whether through individual or collective petition to the management directly in charge, through appeal to a higher authority with the intention of forcing a change in management, or through various types of actions and protests, including those that are meant to mobilise public opinion.'

So the CCC provided a 'voice' — aimed at effecting policy change, at mobilising public opinion — even if sometimes no one was listening. In the absence of that it was difficult to see what other platform would have brought together, say, the solidly Conservative-run (as of 1992) non-metropolitan district council of Gedling in Nottinghamshire, with the Labour, virtual one-party state in the 1980s and early 1990s of Barnsley Metropolitan Borough Council in Yorkshire (see Turner, 1988).

The position of pressure groups in politics is never static. Their fortunes wax and wane — as is argued within the pluralist position in political science (Ham and Hill, 1984; Polsby 1963) — in relation to their political 'strength'. Such 'strength', or the absence of it, may find expression in financial terms; in numerical terms (numbers joining the group: the more, the greater strength); or in terms of the extent of popular sympathy to the cause (on this basis the Child Poverty Action Group was likely to be a lot stronger that the Pædophile Information Exchange); or if the pressure group was a trade union, strength might be measured by its ability to command industrial action which adversely affected the economy. Essentially, pressure groups exist to place the demands of their members on the political agenda; and to exercise pressure designed to get those demands met.

In political science terms, there have been many attempts to impose a categorisation on pressure groups. Stewart (1958), for example, divided pressure groups into 'sectional' (or 'interest') groups and 'promotional' (or 'cause') groups. The former would aim to promote the interests of a section of the community: 'their membership is normally restricted to that section' (Stewart, 1958). Promotional groups, on the other hand, represent a belief or cause — such as the RSPCA or Friends of the Earth — and membership would be, at least theoretically, unrestricted.

Both the NUM and the Coalfield Communities Campaign could be seen as being 'sectional' or interest groups. The trajectory of the former, and the reaction that the latter received, in relation to both their attempts to locate themselves in the policy-making arena, are worthy of brief investigation.

Clearly, following the election of a Labour government in 1974, the NUM moved from being an 'outsider' group under Mr Heath's Conservative government — under which they had fought in one case, and commenced the fight of in the other, and, won, two major national strikes in 1972 and 1974 — to an 'insider' group under the Labour government. An 'insider' group on Grant's (1990) definition, was a group 'consulted regularly by government departments on issues which concern them'. Outsider groups are those which are not consulted by government.

The NUM's 'insider' status under the Labour government of 1974-79 was evidenced by the incorporation of the union — alongside central government representatives, the other mining unions BACM and NACODS, NCB management — in the tripartite negotiations which eventually led to the publication in 1974 by the government of a

68

document promising expansion and investment in the coal industry which became known as 'Plan for Coal' (Department of Energy, 1974). The Labour government also set up the Coal Industry Tripartite Group — of which the NUM, alongside the other two mining unions, the NCB and representatives of central government, were members — to formally approve 'Plan for Coal' and to monitor progress. This was another indication of 'insider' group status of the NUM under the 1974-79 Labour government. Yet another indication of the 'insider' status of the NUM during this period was the attempt by Tony Benn in 1978, then Secretary of State for Energy, to give the NUM veto powers over pit closure decisions (*The Times*, 9 June 1978). The proposal was after protests by the NUM at the NCB's decision to close the Teversal colliery on the Nottinghamshire-Derbyshire border. The existence of the proposal is confirmed in Benn's diaries, which were published later (Benn, 1990) (see later in the text for more details).

On the election of Mrs Thatcher's Conservative government in 1979, or at least certainly since the election of Arthur Scargill as national NUM president in 1981, the NUM became, in Grant's terms, an 'ideological outsider group' (Grant, 1990).

The industry had had a history of consultation since nationalisation between management and union representatives (Scargill and Kahn, 1980) — over closure proposals, or working conditions, for example — and a history of consultation between central government and coal industry management (Hall, 1981). Again, this might be over closure proposals, but also it ranged more broadly over other matters: investment levels, wage levels, over the extent to which the domestic market would be open or closed to imported coal; over relationships and contractual arrangements between the coal industry and the electricity generators and so on. Consultation at industry level — between management and the NUM — continued, in the early part of the 1980s, but there were no consultations at governmental level between the NUM and central government. The first meeting between the NUM and central government after

69

1983 took place in August 1992. At that meeting, the NUM made clear its total opposition to the government's plan to privatise British Coal (*The Miner*, 1992). Eventually, particularly after the 1984/5 coal industry dispute over pit closure proposals, consultations at virtually every level give way to ultimatums and directions from NCB/British Coal management and central government. Within the space of few years, the NUM had moved from being an 'insider' group to an 'ideological outsider group'. As Grant (1990) noted, unlike 'insider' groups, who have to 'adhere to an unwritten code of moderate and responsible behaviour, to informal "rules of the game",' ideological outsider groups face no such constraints, and hence 'have a wide range of strategies open to them'. As evidence to support Grant's first point — 'unwritten code of moderate and responsible behaviour' — it might be noted that there were no national miners' strikes while Labour were in power between 1974 and 1979.

In light of the personal ideological stance of Arthur Scargill himself — which, incontrovertibly, could be called anti-capitalist and socialist (Scargill, 1975) — being in the position of having the pressure group he led as an ideological outsider group, probably suited him in the context of a free market, and fairly right-wing Conservative central government in power. Radical left-wing change in society would not be achieved, obviously, by being in an 'inside' position with a Conservative, free market government. All that could be achieved from that position would be acquiescence with rationalisation and closure proposals: otherwise the 'insider' position would disappear. As an ideological outsider group — if it could wield power which would potentially affect the economy adversely — there was at least a chance that the 'demands' it sought might be met. This, after all, had been the case in the 1972 and 1974 strikes when the NUM had been an 'outsider' group in relation to the Heath Conservative government. Scargill (1975) made his position quite clear, supporting the contention that he would be happy with the status of 'ideological outsider group' for the pressure group he led. At least in the context of a Conservative government in power:

'You will not convince the working class of the need to struggle unless you show them by example, as at Saltley, that struggle is a result of our legitimate claims, and that the system will continue to affect them ... struggles convince workers of the need for real control over society. Saltley and the miners' struggles of '72 and '74, did more to convince the miners in the coalfields of the need to take into social control all the means of producing wealth and not **just** nationalisation, then anything that I know'. (emphasis in text).

In contrast to the NUM's 1980s and early 1990s stance, the Coalfield Communities Campaign was much less hide-bound to an ideologically preconceived position. In Grant's (1990) typology on the national scene in Britain, they represented a 'potential insider group':

'They would like to become insider groups but face the problem of gaining government's attention as a prelude to their being accepted as groups which should be consulted in relation to particular policy areas.'

Grant's (1990) view above would become applicable to the CCC only with qualification, however. Basically, the CCC would have liked to have been an insider pressure group at both national government level and European Community level. They were on their way to it at the latter level, but at the former, for the status to be achieved, it would have required a modified ideology at governmental level. Nevertheless, the **evidence** that insider group status was the objective at both levels of government was provided by the CCC's inclusion of local authorities from all shades of the conventional political spectrum and their orthodox attention-getting techniques of lobbying and seeking to attract publicity. In other words, to employ the Grant (1990) terminology, they were operating within the 'rules of the game.' The NUM were not bound to adhere to the 'rules of the game'; and had, consequently 'a wider range of strategies open to them.' Hence the 1981 dispute and the 1984/85 strike against pit closures.

2.10   Conclusion

The special economic 'needs/demands/wants' of Britain's deindustrialized coalfields achieved only partial acceptance on the British central government's 'policy agenda' in the 1980s and early 1990s. The 'needs/demands/wants' achieved more success in gaining recognition at the level of the Commission of the European Community's 'policy agenda.'

The reasons for this state of affairs can be located in the ideological, and in the fact that the Conservative Party — who were in central government power in Britain for the whole of this period — had little to lose in terms of political support by not making coalfield economic regeneration a priority. At the level of the Commission of the European Communities, more success was achieved in gaining recognition for the economic 'needs/demands/wants' of some of the dwellers of Britain's deindustrialized coalfields on the 'policy agenda' because of an absence of ideological hostility at that level towards these 'needs/demands/wants' and their proponents. The Commission also had a more neutral stance towards political 'support' as it has been defined here. It was removed from the political competition provided by democracy within which British central government had to play. It had no elections to win or lose. Partly as a consequence of that it could be argued that they retained a greater level of pragmatism in relation to the responsible husbandry of the European Community's energy resources. That was a pragmatism which retained at least the potential for a role for state intervention, even if this was the state at supra-national level. The position of the British central government at this period, as was evidenced by privatisation of energy producing, electricity generating and distribution organisations, was the converse of this. The market would prevail,

72

in energy as elsewhere in the economy, and decisions would be decentralised to the level of actions by companies and individuals within the market economy. The role of the state was minimal, if it existed at all.

In relation to one of the pressure groups seeking to have certain coalfield 'demands' recognised and then met — the NUM — that group was viewed by central government as possessing something less than 'legitimacy.' Indeed, in political science terminology it could be classed as being an 'ideological outsider group,' from the perspective of the British central government for the time period under consideration. The other pressure group involved in articulating the 'demands' of some of the dwellers of the coalfields and former coalfields achieved more success in establishing its 'legitimacy,' at least at the level of the CEC. Their decision to lobby at that level indicated that they felt they were more likely, from an ideological perspective, to gain a recognition for their 'needs/demands/wants' at the European Community level than they were at national level. It was also at least partially indicative that, as regards policies on regional development, decision-making during this time period was moving away form the level of the nation state and towards the level of the European Community.

In terms of 'feasibility', there could be little doubt that it would have been technically possible — in relation to the science of applied economics, and in relation to the allocation of greater than was allocated amounts of public expenditure — for the British central government to have placed a greater policy emphasis on the economic regeneration of the coalfields than actually was granted over the period in question. A major conclusion has to be that, in relation to policy emphasis, the British central government, from the mid-1980s to the early 1990s, attached a far greater policy priority to the establishment of a 'market clearing', reduced drastically in size, free of government subsidies, privately-owned coal industry, rather than the economic well-being of coalfield communities. This was in line with its policy priorities in relation to the rest of industry and businesses operating in Britain (Turner, 1989). A second policy priority which took precedence over any interest given to economic regeneration in the coalfields was clearly the perceived need to constrain public expenditure: this in itself, for example, was a primary reason given by the British central government as to why the policy option of closing 'uneconomic pits' had to be pursued in the first place. Keeping open 'uneconomic pits' meant subsidies from government, which raised public expenditure, which in turn fed through into the demands the government had to make on taxpayers. High taxation acted as a disincentive to individuals to work, and therefore militated against generalised economic progress (Turner, 1990).

Coal mining was unique in Britain in the extent to which the communities it spawned developed their own distinctive social culture, and unique in the intensity of that culture. It was a culture which emphasised the community above individualism; it was a culture which provided for inter-generational reproduction of labour for generally unpleasant tasks (Allen, 1981), and provided for inter-generational skills transference (son followed father down the pit); it was a culture born out of periodic traumatisation by terrible accidents and sacrifices above and below ground. The culture was continually influenced by generally unpleasant working conditions, and job insecurity even after nationalisation in 1947. A strong case could be made for arguing that in respect of all this, even if there were sound economic arguments for pit closures, the 'need' for economic regeneration measures in the coalfields deserved a more central place on the government's policy agenda than they actually achieved in the 1980s and early 1990s, at least prior to October 1992.

3   Policy making for regeneration in the coalfields: A changing framework

3.1   From regional policy to local projects

The 'traditional' policy mechanism in
Britain by which attempts have been made to
alleviate 'territorially-specific industrial
decline' (Morgan, 1985) — or failure to develop
a buoyant local or regional economy in the first
place — has been regional policy. It is the
oldest form of what later came to be termed
'industrial policy,' dating back to the Special
Areas Act of 1934 (Hogwood, 1987; Turner, 1989).
To start with, the building of factories and
industrial estates was the central part of the
policy; only later were incentives provided to
firms to locate themselves in depressed areas in
the form of loans, grants and tax relief.
    Regional policy eventually had two major
constituents: sanctions   and   financial
inducements. The sanction part was essentially
that companies would be refused permission,
through a form of planning control (the
withholding of Industrial Development

Certificates or Office Development Permits) from locating or expanding in areas that were relatively prosperous, such as the West Midlands or South East of England. Rootes, for example, the British vehicle manufacturer that was later taken over by the US multi-national Chrysler, was refused permission to expand in Coventry — at that time the heart of the British motor manufacturing industry — and ended up siting its new factory in 1963 in Linwood in Scotland (Wilks, 1984). The second part of regional policy was provided by a mixture of automatic and selective (until 1988, when all grants became selective [Department of Trade and Industry, 1988]) financial inducements to get firms to move to areas of high unemployment. Sometimes, for example, the government would organise and finance the building of a factory itself; sometimes it would provide capital grants; another example of regional policy in action was the payment of Regional Employment Premium to firms between 1967 and 1977. In one way, this latter example was an exception: it was a labour subsidy, while most of the regional policy inducements were subsidies for capital (Hogwood, 1987). Such is the change in economic times between the 1960s and the 1980s, that when in 1960, the then British Motor Corporation announced it would be opening not one but **three** new factories, with regional policy assistance, in areas of high unemployment (South Wales, Merseyside, Scotland), not only was this **not** front page news but it failed to better than be reported as just another item on page 8 of The Times! (*The Times*, 22 January 1960).

In the context of regional policy, the history of attempts to regenerate declining coalfields is a long one. Heughan (1953) noted that regional policy was in operation in a survey undertaken in 1949 of the then declining coal mining locality of Shotts, a collection of former pit villages between Edinburgh and Glasgow. She noted how employment prospects for female labour were being improved:

'By January 1949 the building of a new and much larger factory was almost complete. This was being built under the Government's

76

industrial development scheme and was to house a branch of Summer's (Wren Nest) Cotton and Rayon factory from Gosport. '

Similarly, in November 1967, the Labour government announced the creation of Special Development Areas — where companies would be eligible for enhanced assistance — within areas that were already eligible for assistance under regional policy. To start with, these Special Development Areas were mainly areas facing declining coal mining such as the North East and South Wales (Hogwood, 1987).

Attempts at regenerating coalfield areas in the 1980s and early 1990s were, however, taking place within the context of a declining emphasis on 'blanket' regional policy which would cover entire areas (Morgan, 1985; Turner, 1989). The 1983 White Paper, *Regional Industrial Development*, abolished Special Development Areas, fixed new cost-per-job limits on assistance to firms, reduced the geographical coverage of Development Areas — where companies were entitled to financial assistance automatically — and increased the geographical coverage of Intermediate Areas, where companies were eligible only for selective, or discretionary, assistance. Monies provided for regional preferential assistance fell from £1011.9 million in 1982/83 to £610.0 million in 1984/85 (at 1984/85 prices), (Hogwood, 1987). Industrial Development Certificates were formally suspended as a policy tool in 1982, though had not been strictly enforced since the 1960s (Morgan, 1985).

The 1988 White Paper referred to earlier, DTI - *the Department for Enterprise*, further scaled down regional policy by ending all automatic subsidies to companies. So, regional policy generally, scaled down in 1983 and 1988, gave way to the creation of Urban Development Corporations, inner city Task Forces, enterprise zones, and to the City Challenge schemes as new local project-based policy mechanisms aimed at regeneration. Policy for regenerating the coalfields and former coalfields was no exception to this policy fragmentation and policy modernisation affecting government action targeted at economic reconversion and renewal. Thus, for example, on the same day in November 1984, as the Department of Trade and Industry announced that financial assistance to companies under regional policy was to be cut in half — from £660m per annum to £300m — the Department of Energy announced it was doubling the amount of public funds being made available to NCB Enterprises (the job creation arm of British Coal, which later became British Coal Enterprise) to £10m per annum (*The Guardian*, 29 November 1984).

No matter what the ideological hostility towards it from central government at this time, some regional policy had to remain in place at the national level, in order to continue to attract funds from the European Community for regional development (Morgan, 1985).

This policy modification outlined above reflected the Conservative central government's emphasis on a more **market-orientated** stance towards policy aimed at initiating economic renewal, and a commitment to **target** policy more accurately at intended recipients or intended localities. There was a parallel here with how the Conservative central government of this era handled its approach to what was at one time called 'welfare state' policy: the Conservatives were concerned to 'target' benefits to those deemed as deserving them and modified benefit policy accordingly, moving away from blanket, universal coverage. A robust regional policy — as pursued in the 1960s and 1970s — would have been ideologically unacceptable to the Conservative central government in the 1980s and early 1990s. It would have been seen as an unnecessary, and even damaging, government interference in the management of companies (Turner, 1989). There was also a certain pragmatism involved in the policy modification away from blanket regional policy and towards the more carefully targeted local projects, however. After all, regional policy could present the potential prospect of effecting a major impact on regional economies **only** if a national economy was in a sufficiently buoyant condition such that firms **were** actively expanding and relocating. Though in certain

industrial sectors in the 1960s that **was** the
case, it was certainly not the case in the early
1990s, and for large parts of the 1980s. During
the early 1990s, the British national economy
was gripped by what many termed a 'recession,'
though the roots of that 'recession' might have
been deeper than the temporary fall-off in
aggregate demand that the term implied. The
economic condition of the early 1990s might well
have been a consequence of a **structural** change
in the economy brought about by the deindustrialization discussed in more depth in
chapter one, compounded by recessionary factors.

Where other agencies — such as local
authorities — entered the regeneration
business, this was at least partially because
they perceived a lack of central government
input, or at least a lack of central government
priority, towards economic regeneration efforts.
This was coupled with a generally-held view
within relevant local authorities that they
could not just sit back and do nothing while
their local economies deindustrialized.

It is important to recognise that the
creation of paid employment opportunities was
not, for many years, the central policy
objective of regional policy. The attempt at
direct linkage between job creation and regional
policy was abandoned with the abolition of the
Regional Employment Premium, mentioned earlier,
by the Labour government in 1976. Instead, the
central objective of regional policy was more
usually to increase **investment**, hence its
emphasis on capital incentives to firms (Morgan,
1985). By 1983, the Conservative central
government was arguing that the primary case for
maintaining regional policy was social, rather
than economic. Regional policy henceforth was to
have a 'single guiding objective, that objective
being to reduce regional disparities in
employment opportunities on a stable long terms
basis' (DTI 1983).

It has been recognised by many that the
formerly assumed direct link between investment
and job creation no longer necessarily existed
from about the mid-1960s onwards. André Gorz
(1985), for example, referred to research in

West Germany published in 1976 which argued that:

> 'DM1,000 million invested in industrial plant would have generated two million jobs from 1955-60 and 400,000 jobs from 1960-65. From 1965-70 the same sum would have **destroyed** 500,000' (emphasis in text).

Clearly, the pressing need in the coalfields in the late 1980s and early 1990s was for the creation of paid employment opportunities **and** investment, or, alternatively, investment that created paid employment. Sometimes that need coincided with the objectives of the territorially specific regeneration initiatives examined within. In any case, whether or not there was such a co-incidence, the intention in this work was to examine on a case-by-case basis the effectiveness and the *raison d'etre* of a series of economic regeneration projects in the coalfields and former coalfields.

3.2 Policy communities and policy networks

In the late 1980s and early 1990s particularly, though in Heclo and Wildavsky's (1981) case before this, much work on policy making studies focused on the inter-related concepts of policy community, policy network on their concomitant influence on the earlier discussed policy agenda (see, for example, Atkinson and Coleman, 1992; Grant, Paterson and Whitson, 1988; Heclo and Wildavsky, 1981; Jordan, 1990; Rhodes 1986; Wilks and Wright 1987; Wright 1988).

Such a model of policy-making was analytically useful in the sense that it sought to identify who — which groups or individuals — had power in a society and what, if anything, they did with that power. So this approach to policy-making focused on one of the most important areas in policy studies: who gained, and who lost, from different actions by actors in the policy community.

There was no one agreed definition of 'policy community'. One that might have found general acceptance, however, was that the policy community was comprised of those individuals who had the power to take effective action that would influence policy in some particular direction. The most likely candidates for membership, therefore, would be politicians and bureaucrats. Heclo and Wildavsky (1981), for example, argued that the 'community' and 'policy', and therefore 'policy community' could be defined as follows:

> 'Community refers to the personal relationships between major political and administrative actors — sometimes in conflict, often in agreement, but always in touch and operating within a shared framework. Community is the cohesive and orientating bond underlying any particular issue. Policy is governmental action affecting some end outside itself. There is no escaping the tension between policy and community, between adapting actions and maintaining relationships, between governing now and preserving the possibility of governing later. To cope with the world outside without destroying the under- standings their common life requires — this is the underlying dilemma facing the communities of political administrators.'

Heclo and Wildavsky (1981), then, argued that the 'policy community' idea was one way of approaching an understanding of how particular policies were formulated. They recognised the potential for tension as well as agreement amongst the members of the 'community.' And they recognised also the axiom that not only would this community be concerned about preserving its hold on governmental power in the present, but also maintaining and reproducing its hold on governmental power in the future. Policy 'networks' in this scheme of understanding are the interaction between different individuals in the policy community and between different policy communities. In acknowledgement that this area is definitionally disputatious, it should be noted that for Rhodes (1986), the concept of policy network took precedence over policy community: the community was best understood as a network which had a restricted membership, stable relationships between community members, and insulation from other institutions and networks. For Wright (1988), however, the concepts of policy community and policy network were separate. The former included all those who shared a common identity or interest in a policy *sub*-sector (say the drop forgings sector of the steel industry); network referred to the link between different policy communities. Wright (1988) introduced another term — policy universe — which was a much broader concept, and included a much wider range of institutions which may have at some time some interest in a particular policy area. These might include 'Parliament and its committees, political parties, the media, analysts and commentators' (Wright 1988). So, if reference could be made to a 'drop forgings policy community,' reference could also be made to a '(steel) industry policy universe' which would encompass it. Notwithstanding the definitional disputes, the central point was that those wielding influence within the policy community and/or policy network will have an input into what will be prioritised on the already discussed policy agenda.

So, for example, Smith (1989) argued that the existence, and the powers wielded by, an agricultural 'policy community,' prevented the Attlee post-war Labour government from adhering to a long-standing policy commitment to nationalising land. Heclo and Wildavsky (1981) argued that a community of administrators and politicians controlled public expenditure decisions in Britain.

What should be noted is that there would be different communities for different policy sectors: the health policy community, for example, would not be the same as the education policy community. And, as Wright (1988) noted, unless these were broken down into sub-sectional levels — such as schools education policy community, higher education policy community — they were too broad to be of anything other than general application. In relation to the membership of these communities there seemed to be a strong case for arguing that there would be

82

an overlap: the Secretary of State for Education, for example, would be a member of both the schools education policy community and the higher education policy community.

The argument presented herein, however, is that whilst the undifferentiated concept of policy community might indeed be valid in analyzing certain substantive policy areas — agriculture, public expenditure, health, education probably being good examples — in other areas it was not so valid or, at least, it was not valid without considerable qualification. That was a qualification which would become necessary as empirical investigation progressed.

One particular policy area where the concepts of policy community and policy network did not sit so easily was with the variety of attempts at economic regeneration policy in Britain's coalfields pursued in the 1980s and 1990s. That they did not sit easily there would indicate that there might have been other policy areas in which the concepts were similarly inapplicable. If that was so, the temptation to automatically look for 'policy communities' as an aid to policy analysis would, legitimately, be weakened.

Any attempt to apply the policy community/ policy network model as a mechanism for the understanding of policy making in relation to economic regeneration of the coalfields, would reveal no direct fit, for more than one reason. Firstly, that was because of the multifarious nature of the agencies involved in economic regeneration. They included local enterprise agencies; local authorities at three different levels: non-metropolitan county council, non-metropolitan district council, metropolitan district council; central government operating at the local territorial level via an enterprise zone; the Rural Development Commission; the private sector, such as for example with Costain's 'partnership' with Barnsley Metropolitan Borough Council (see earlier in the text); the Department of the Environment via the 'City Challenge' schemes; a civil servant-staffed Task Force in Doncaster; the voluntary, charitable sector. One example of the latter would have been Conservation Practice Limited, based in the Dearne Valley, South Yorkshire, which had as much of a 'social' role as an economic: resocialising the long term unemployed back into the world of work by introducing them to limited-life conservation projects and industrial/construction work.

Secondly, the extensive nature — in territorial terms — of pit closures in the 1980s and 1990s, meant that the multiplicity of agencies being brought forth to respond to the problems have been greater still.

Really, what could be seen in relation to regeneration policy was a collection of different policy experiments sanctioned at different administrative levels. Thus essentially, the end product — the policy output — was a series of ad hoc policy responses to a similar problem of de-industrialization manifesting itself in different geographical locations.

So even if a 'policy community' was evident in particular localities at particular times, it could not be argued that there was an all embracing 'policy community' covering all attempts at regeneration, at least prior to October 1992. It was possible, however, that certain members or components might be shared: for example, a Department of Environment official might be a member of more than one 'policy community' in relation to regenerating the coalfields.

Take the two examples of regeneration examined within which relate to Doncaster. One 'community' that might have been identified was that around Doncaster Task Force, between 1987 and 1990. Such a 'community' might have contained the civil servant members of the Task Force itself, the central government minister/civil servant overseeing the project, and possibly a representative of the private sector. In this case, because Doncaster Task Force were seeking to 'rejuvenate' the local Chamber of Commerce and Industry, the private sector representative might have been the chief executive of that body. A second 'community' involved in regeneration in the same town during a similar time period and beyond would be that located around British Coal Enterprise. Here, the 'community' members might be identified as including local and national representatives of British Coal Enterprise, representatives of British Coal (of which BCE was a subsidiary at the time of writing), and representatives of central government (which, at the time of writing, with British Coal and BCE still in the public sector, determined the budget for both organisations). Information received from interviews carried out for this study suggested that there was very little — if any — interaction between the two 'communities,' and no co-ordination at all despite the fact that they were engaged in the same job (regeneration attempts), in the same town.

After the October 1992 announcement of the closure or mothballing of 31 collieries, it was clear that a central government-led policy community on economic regeneration in the coalfields did emerge. And it emerged in response to the political uproar over the closure decisions.

Where a policy network, as defined above, could be identified, Wright (1988) has argued:

'The evidence, so far, suggests that access to a policy network is normally controlled by the dominant government agency .... [operating within it].'

This may be true. In the case of the micro-industrial policy of attempts at regeneration of the coalfields, however, it seemed in need of qualification. There was, in many coalfield areas, a **dominant** government agency, although in some areas government agencies vied with each other for dominance. The two government agencies mentioned above in Doncaster might be one example of that, in the sense that both were doing a similar job, at a similar time, in the same place and both, presumably, would have liked to have been the recipient of more government money and 'lead' status. Another example might have been the Rural Development Commission and British Coal Enterprise in Bolsover (see chapter four). Again, both were offering similar services in the same place to much the same people/firms, and both were government agencies. Neither of them seemed to be 'dominant' in relation to the other, however. Importantly, where they existed, the dominant agency or agencies might have been actively seeking organisations and individuals to join their 'policy community': there may have been a shortage of skilled professionals and affiliated (unofficially) organisations. Evidence suggested that his was the case, particularly in relation to privately-organised but often partially public-funded local enterprise agencies that could not really offer a high-powered career to a manager. Another example of this is noted in chapter seven: the fact that ICI seconded a manager to Doncaster Chamber of Commerce and Industry might be seen as evidence that the latter body had a shortage of skilled professionals.

One is left with the feeling that the idea of policy communities and policy networks were most applicable to policy areas and organisations which were traditionally easy to delineate and identify: such as industrial policy communities, education policy communities. In, for example, the chemical industry, Grant, Paterson and Whitson (1988), identified five discrete policy communities (Wright, 1988).

Where did this leave the small policy communities that were involved in the micro-industrial policy strategy of regenerating the coalfields? At least prior to October 1992, the picture was one of fluidity and transience, with periodically changing actors. The Task Force in Doncaster, for example, left behind it a more spectral, less accurately targeted collection of policy communities, such as the inchoately regenerated Doncaster Chamber of Commerce and Industry, and the Barnsley and Doncaster Training and Enterprise Council. Similarly, the Bolsover Enterprise Agency (see chapter 4) would continue to exist — therefore continued to be part of 'policy community' — for as long as its finances continued to come in from local government.

The policy area under investigation straddled the sectoral and sub-sectoral levels of industrial policy. The sectoral level of industrial policy which formed the backcloth to, and the context of, this study, was policy towards the coal industry. Here policy was determined at central government level and implemented by British Coal, while the company was in the public sector. Of course, there may have been occasions when that implementation was challenged by other forces. Where such a challenge was successful, that would have represented an unsuccessful attempt at policy implementation by the relevant authorities. The NCB management/Conservative central government, for instance, failed in 1972 and 1974 to impose their chosen policy on wage levels for coal industry employees, as the wage levels were successfully resisted in two national strikes.

At the sub-sectoral level, this investigation was centrally concerned with economic regeneration policies carried out within circumscribed territorial locations. These could be referred to as 'pit closure zones.' It was here that policy objectives in relation to economic regeneration — which may have been set at central government level, local government level, or at the level of some other organisation aiming for economic regeneration — merged into behaviour by key actors as the policy determinant. Considerable autonomy was discovered in relation to the levels of discretion wielded by local officers of, say, Bolsover Enterprise Agency (see chapter 4), for example; British Coal Enterprise (see chapter 5); the Doncaster Task Force (see chapter 7); the Barnsley Business and Innovation Centre (see chapter 8). As Wright noted:

'Policy is influenced, shaped and sometimes determined behaviourally by those responsible for its initiation and formation, as well as by those responsible for carrying it out. Behaviour may be the policy.'

3.3   Conclusion

The central policy mechanism applicable to regeneration in general, and coalfield regeneration in particular, moved in the 1980s away from being reliant on a 'blanket' regional policy approach, and towards a series of more specifically-targeted territorially-specific projects.

As a conceptual tool to aid the understanding of policy-making for regeneration in the coalfields, the idea of 'policy communities' and 'policy networks' needed substantial qualification and could offer only a partial understanding that formed an adjunct to the ideas on 'policy agenda' discussed in chapter 2. The idea of a 'policy community' operating was probably a lot stronger in relation to the **contraction** of British Coal, rather than in relation to the regeneration efforts that followed that decline, at least prior to October 1992. The policy community that could be identified in relation to contraction included ministers and civil servants at the Department of Energy and, after the DoE's abolition, the Department of Trade and Industry, alongside British Coal management.

The nuts and bolts of policy activity in relation to regeneration — which strategy to adopt, which companies/individuals to help, and why — often boiled down to the autonomous and discretionary actions of local officials of regeneration agencies. In other words, this is a study which largely focuses on the behavourial actions of local officials rather than on any centralised, macro-policy response to effect regeneration. The former existed, and it can be argued that the latter did not until October 1992. The evidence to support that assertion runs through this text.

4 Deindustrialization and reindustrialization in Bolsover: A multi-agency response to economic change in the North Derbyshire coalfield

4.1 The locality

Bolsover is a bastardized administrative unit, as of 1992, consisting of a collection of small towns in the county of Derbyshire. Unlike many local government administrative units, it did not have a large town as a central focus. Instead, the town of Bolsover, with a population of 11,320 as of mid-1989, provided its central core with other population centres being the smaller, adjacent towns of Shirebrook, South Normanton, Clowne and Scarcliffe. What did link the towns for many years, however, was the dominance of the coal industry in the local economy. It was home to the largest coke preparation plant in western Europe (Municipal Year Book, 1991), operated by the private sector company Coalite, employing, as of July 1991, 892. Adjacent to this was a chemicals-from-coal plant, operated by the same company, employing 375 during the same period. At one side of the Coalite plants, in 1991, was Bolsover colliery.

At the other, as of the same date, was Markham colliery. A few miles up the road was Shirebrook colliery. In 1981, 26 per cent of the male population in Bolsover district and the adjacent local government administrative unit of North East Derbyshire were employed in the coal industry (North Derbyshire Coalfield Partnership, 1990). In Bolsover itself, the figure was higher: 46 per cent of males were employed in the coal industry (Coalfield Communities Campaign, 1986). Like other coal-producing areas in Britain, the Bolsover district and the region surrounding it suffered a rapid run down in its basic industry, dating from the late 1960s and intensifying in the 1980s. In addition to the loss of white-collar jobs in 1987 with the closure of British Coal's north Derbyshire administrative head-quarters in Bolsover, the Bolsover district — defined as a local government territorial unit — saw Glapwell and Langwith collieries close in the 1970s. In 1986, Whitwell colliery, also in Bolsover district, closed with the loss of 750 jobs (*The Guardian*, 4 June 1986). Elsewhere there was partial closure or rationalisation involving redundancies. Markham colliery, for example, across the border in North East Derbyshire District, which had employed 2164 in 1985, had reduced its workforce to 1380 by 1991. Shirebrook colliery, inside Bolsover district, which had employed 2044 in 1985, employed only 1280 in 1991. Between 1968 and 1991, 15 pits within a radius of five miles of Bolsover town were closed or merged, involving a subtraction form that local economy of 17,647 jobs in the coal industry (see Table 1).

This investigation concerned itself with economic regeneration policy as it stood in 1991 in this heavily deindustrialized locality: the strategy, the policy-creating actors, and the objects at whom the policy was targeted. It is based on a face-to-face interview with a leading official of Bolsover Enterprise Agency, and telephone interviews with the owners/managers of 28 companies on Bolsover Enterprise Park. All the interviews took place between May and July 1991. First, however, to enable an understanding of the nature of the local economy, and

therefore of the local economic context in which
regeneration efforts were taking place, it would
be worthwhile to briefly sketch the pattern of
deindustrialization here.

Table 1

Coal Mines within five miles of Bolsover town 1968 onwards

	1968 Labour Force	1985 Labour Force	1991 Labour Force (a)
Arkwright	776	651	N/A Closed 1988
'A'Winning	1,023	N/A Closed 1969	
Bolsover	891	940	528
Cresswell (b)	1,191	1,113	N/A Closed 1991
Glapwell	1,864	N/A Closed 1974	
High Moor	363	587	241 (c)
Holmewood	1,257	N/A Closed 1968	
Ireland	801	707	N/A Merged with Markham – 1987
Langwith	1,143	N/A Closed 1978	
Markham	2,088	2,164	1,380
Oxcroft	745	N/A Closed 1974	
Pleasley	1,102	N/A Closed 1983	
Renishaw Park	479	567	N/A Closed 1989(c)
Shirebrook	1,460	2,044	1,280
Silverhill (b)	1,130	1,108	781
Teversal (b)	842	N/A Closed 1978	
Warsop (b)	1,309	1,234	N/A Closed 1989 (d)
Westthorpe	804	N/A Closed 1984 (c)	
Whitwell	842	771	N/A Closed 1986 (e)
Williamthorpe	1,747	N/A Closed 1970	
TOTALS	21,857	11,895	4,210

Notes to Table 1

(a) Except for Silverhill, information here is as of 27
     July 1991, and data is from Derbyshire County
     Council, Planning Department. Silverhill figures are
     for 1990 and are from *Nottingham Evening Post*, 8 May
     1990.

(b) Nottinghamshire administrative unit of British Coal,
     but within 5 miles of Bolsover town. Cresswell
     colliery, geographically, was in Derbyshire.

(c) Merged with Kiveton Park colliery (Yorkshire)
     December 1991, but remained a work base for 241 (see
     North Derbyshire Coalfield Partnership, 1991).

(d) *The Times*, 29 June 1989.

(e) *The Times*, 4 June 1986

Sources other than those named above: NUM Derbyshire area;
Guide to the Coalfields, 1968, 1969, 1971, Colliery
Guardian, London; *Guide to the Coalfields*, 1986, 1990,
Colliery Guardian, Redhill, Surrey.

4.2   Deindustrialization

The 'politics' of deindustrialization had a
special importance in this coalfield locality.
Some of the earliest attempts by the National
Union of Mineworkers to stop pit closures
centred on collieries here. A few miles from
Bolsover colliery used to lie the north
Derbyshire Langwith colliery, for example.
   In February 1976, the national executive of
the National Union of Mineworkers imposed a
national overtime ban in protest against the
planned closure of Langwith. According to the
then National Coal Board, its seams had become
too expensive to mine: it was an 'uneconomic'
pit of the 1970s (*The Times* 4, 13 February
1976).
   The significance of this was that it was the
first time in the history of the NUM that
national industrial action had been taken with
objective of saving a pit from closure. In that
sense, in relation to the arguments advanced in
Chapter 2, it can be argued that this was the
first time that the NUM had attempted to place
this 'demand' — that a coal mine should not
close — on the national political agenda. A

minority, but developing, left-wing faction had been arguing since the 1960s that the NUM should take a stand against pit closures. The established position amongst both left and right in the Union leadership since the national-isation of the coal industry in 1947, however, had been to oppose the idea of taking industrial action against pit closures. Questions on the size of the coal industry were considered by them to be outside the scope of the Union's industrial ambit. The NUM should, instead, concern itself with wages, working conditions, compensation claims. Issues of the size of the industry were issues for the political wing of the Labour movement: the Labour Party and Labour governments. Following nationalisation, it was considered by the established leadership of the NUM — until the changes in that leadership in the 1980s — that there were 'no opposing sides in the industry,' and that taking industrial action against pit closures would have been unnecessarily disruptive (Turner, 1985).

So, for a long period, pit closures were taking place in a de-politicised context. Not only were the problems caused by closures, and the energy requirement arguments against pit closures not on the political agenda as it was defined earlier, nobody, including the NUM, had tried to put them on.

Nevertheless, there was a certain irony attached to the fact that Langwith should have the honour bestowed upon it of being the centre of the NUM's first national fight against a pit closure. Many of those still employed in the coal industry in the locality during the 1984/85 strike against pit closures opposed that latter attempt to stop deindustrialization; that latter attempt to gain prominence on the national political agenda for the issue of, and the problems of, pit closures. At Bolsover colliery itself, for example, workers voted by 479 to 341 in 1984 against taking industrial action to try to stop closures (Morgan and Coates, undated). In November 1984, eight months into the NUM's strike, the National Coal Board reported that virtually half of the workers at Shirebrook colliery had returned (*The Guardian*, 15 November 1984). By January 1985, the NCB was publicising

94

its claim that 5,089 — or 48.5 per cent — of the total north Derbyshire coalfield labour force of 10,500 had returned to work (*Financial Times*, 8 January 1985). This contrasted with a much more solid response to the strike at this stage in its progress from those employed in the industry in the South Yorkshire (see Winterton and Winterton, 1989) or the South Wales coalfield, for example.

The claims of the NCB regarding coal industry employees returning to work during the 1984/85 strike had to be treated with caution, obviously. The NCB, for its part, was seeking to maximise the claimed numbers of miners returned to work as part of the propaganda battle with the NUM. Moreover, returning to work was not necessarily evidence of being against the fight to stop pit closures, the fight to get the issue of pit closures on to the political agenda. It could, instead, be a consequence of economic hardship. Nevertheless, there remained evidence that many coal industry employees in this area were against taking industrial action to stop pits from closing in the 1980s.

And, in relation to the overtime ban over Langwith mentioned above, it should be noted that there was strong opposition to this industrial action from many individual members of the NUM, and some branch and area unions; opposition that can be construed as opposition to the NUM trying to place the issue of pit closures on to the political agenda: opposition to the 'politicisation' of the issue. Joe Gormley, the then President of the NUM, was clearly annoyed by the decision to impose the ban. From his perspective, the dichotomy in the Labour movement between its 'industrial' and 'political' wings and their relevant ambits remained sacrosanct; and the imposition of this overtime ban violated that separation. He commented on the decision:

'I was surprised by the vote. I must admit .. I am led to the opinion that it became a political decision. There seems to be no other reason' (*The Times*, 13 February 1976).

And later, by which time Gormley was making less attempt to conceal his anger:

'I am browned off and disgusted. I think it is a fiasco' (*The Times*, 13 February 1976).

Seven days later, Mr Gormley managed to persuade the national executive committee to recommend that the ban be called off (*The Times*, 20 February 1976). In a ballot to endorse the national executive's decision, 109,410 mineworkers supported the recommendation, with 69,369 against (*The Times*, 9 March 1976). The first serious attempt to try to get the NUM to use its influence to force the issue of pit closures on to the national political agenda had failed.

Prior to the lifting of the ban, the left-wing leadership of the Yorkshire area of the NUM — by this time, under the Presidency of Arthur Scargill, committed to fighting pit closures by virtually any available means, and who saw pit closures as an issue of national as well as local importance (Turner, 1985) — had circulated thousands of copies of a leaflet which urged the Yorkshire miners to vote against the lifting of the overtime ban. The leaflet claimed that if the NCB could close Langwith, a profitable pit with adequate coal reserves, 'then God help half the pits in Yorkshire.' A 'no' vote would mean that the NEC (of the NUM) and the Coal Board would have to rethink their attitude in respect of Langwith and on the question of pit closures (*The Times*, 1 March 1976).

And a 'no' vote, of course, would have gone some way towards placing the issue of pit closures on to the political agenda — as defined earlier — even if nothing was done in response to that by government or quasi-governmental (in this case the NCB) agencies.

Langwith was closed in 1978. Those within the NUM arguing for industrial action with the objective of stopping pit closures had not reached a large enough number, nor persuaded enough powerful members within the union, to enable a national dispute over pit closures to be prosecuted. Though there were majorities in the ballot for continuing the overtime ban in Kent, South Wales, Scotland and Yorkshire, overall the men could not be persuaded to take action designed to save Langwith. Even in north Derbyshire itself a majority of members were opposed to continuing the ban (Allen, 1981).

There were other collieries that were closed during the late 1970s on the grounds that they were unprofitable, or that new reserves would have been too expensive to develop. Rockingham colliery, near Barnsley, was closed on these grounds in 1977, for example (Turner, 1985). So was Teversal in 1978 — which, as Tevershall, had provided both a pit and a pit village to symbolise the ugliness of industrialisation in D.H. Lawrence's *Lady Chatterley's Lover* — a pit within a five mile radius of Bolsover town. (Allen, 1981). The men at the latter colliery continued the overtime ban discussed above after the national ballot had overturned it, but, in the absence of both national and area support, the protest came to nothing. In the NUM's Nottinghamshire area, 72 per cent of miners voted in early 1979 against taking industrial action in support of saving Teversal (Allen, 1981 p.303). In the five years up to April 1980, the National Coal Board closed twenty-seven collieries across the country, which had provided 14,000 jobs for miners (Hall, 1981 p.254).

It was the protests to central government by the NUM over the intended closure of Teversal, on the Nottinghamshire/Derbyshire border, that led to the first and only offer by government to involve the NUM in the management of the size of the industry. It could be argued that this was the first time the issue, and problems of, pit closures had reached the national political agenda, even if that achievement might be characterised as being tangential and transient.

Tony Benn, made Secretary of State for Energy in June 1975, offered the NUM veto powers in 1978 over any NCB pit closure proposal (*The Times*, 9 June 1978). Effectively, Mr Benn, whose proposal was made apparently in private talks between himself and the NUM's national officials, was offering to abrogate the NCB's prerogative to manage the industry as they saw fit.

Benn's move was confirmed in his published diaries:

'I had told Mick McGahey privately that I intended to inform the Coal Board that they could not close pits without the miners' consent ...

... the only way to achieve this was for me to instruct the Coal Board that closures must be agreed with the NUM' (Benn, 1990,p.295).

By way of an illustration of a point of contrast between that Secretary of State for Energy and a later successor, it was the 'management's right to manage' that was to figure prominently on the Thatcher government's and NCB management's agenda as a central point of non-negotiation during the 1984/85 coal industry dispute.

In what in retrospect seems an extraordinary decision, given the circumstances of the 1980's, the NUM rejected Mr Benn's offer. Joe Gormley, the President; Lawrence Daly, the General Secretary; Mick McGahey, the Vice-President; all counselled against the acceptance of the proposal. Peter Heathfield, then General Secretary of the North Derbyshire area of the NUM, was also critical of the proposal. Heathfield argued that if the Union had a veto it would be blamed for closures which were in fact the NCB's responsibility. On the national executive of the NUM, apparently only Arthur Scargill was enthusiastic about the idea.

What emerged as being crucial to a full appreciation of the deindustrialization-regeneration debate, the impact of pit closures and the arguments against pit closures, was the evidence from the north Derbyshire coalfield on the long term consequences of colliery closure. Male unemployment in Whaley Thorns, within Bolsover district, for example, where the Langwith colliery was situated, stood at 15.2 per cent in December 1990, more than twelve years after the closure of the pit (North

Derbyshire Coalfield Partnership, 1990). This compared with 8.4 per cent for the male average nationwide in December 1990 (Department of Employment, 1991). At Holmewood, across the border in North East Derbyshire, where Holmewood colliery closed in 1968 and Williamthorpe colliery closed in 1968, male unemployment was 14.7 per cent in December 1990 (North Derbyshire Coalfield Partnership, 1990). This was despite the development of 60 acre industrial estate started in 1973 on the site of the former Holmewood colliery and another 53 acre industrial estate started in 1985 across the road on the site of the former Williamthorpe colliery.

4.3 Regeneration

Bolsover Enterprise Park was established in March 1988, on the 12.5 acre site adjacent to Bolsover colliery vacated when British Coal closed its north Derbyshire administrative headquarters in 1987. Those headquarters had employed 300, some of whom transferred to British Coal's Leicestershire offices. The reorganisation took place upon the merger of British Coal's north Derbyshire administrative unit with its south Midlands administrative unit (*Financial Times*, 27 September 1986). The south Midlands administrative unit had covered the territorial areas of south Derbyshire, Leicestershire and Warwickshire and, prior to its total demise in 1989 with the closure of the Betteshanger colliery (*The Times*, 23 August 1989), the three pits of the Kent coalfield.

The Enterprise Park was administrated by Bolsover Enterprise Agency, a small scale operation which, as of July 1991, had three part-time employees, one full-time employee and one caretaker. The Enterprise Agency itself was set-up on the initiative of Bolsover District Council, who were understandably concerned about the run down in the local coal industry.

The Enterprise Park was the first, and at the time of writing, the major, initiative of the Enterprise Agency. As of July 1991, the recurring total annual budget of the latter was £150,000, of which £30,000 per year came from

Bolsover District Council, and £2,000 per year from British Coal Enterprise (*The Guardian*, 5 August 1991); the remainder of the sum being collected as rent from Enterprise Park tenants. Demands on this £150,000 ranged from rent for the land which, despite its former public sector coal industry use, was in private sector owner-ship, wages and salaries of the Enterprise Agency's staff, through to stationary and the upkeep of the Enterprise Agency's office.

The regeneration ethos adopted by Bolosver Enterprise Agency was to provide a physical and economic (in other words, not excessive levels of rent) environment within which local businesses could establish themselves and (hopefully) expand. All but one of the 36 business tenants (see Table 2) of July 1991 were small businesses: the exception was a Chambourcy (Nestlé) distribution depot. The one other occupier, apart from the Enterprise Agency staff themselves, was a training centre run by the local Training and Enterprise Council Centre, which was operating in a building used formerly by Nacro.

Essentially, the activities of the Enterprise Agency could be summarised as follows: firstly, they acted as providers of industrial property to businesses. The attempt was to make this available at as cheap a rate as was possible: £3 per square foot as of July 1991, which would compare with average industrial rents of £3.75 per square foot in another 'pit closure zone' — Barnsley — in 1991 (Brown, 1991a), or £7 per square foot at the same period in one of the relatively prosperous parts of the country, Milton Keynes (Brown, 1991b).

A positive aspect of this was that they were able to recycle a collection of former NCB buildings ranging from office accommodation to out-buildings. The Agency believed that the ad hoc rather than purpose-built nature of this industrial accommodation had actually encouraged a greater variety of businesses to locate there; one-man or one-woman businesses were able to find a 'corner' of the building from which to operate, and might not have been in that position had this been a conventional, purpose-built industrial estate. The operating ethos of the Agency was 'easy on and easy off': the tenants were given two year leases on property but, in practice, the small business owners were never held to pay the full lease if their business folded prior to the expiry of the period.

In relation to regeneration, a general locality level of low rents can have both positive and negative aspects, however. Certainly they were beneficial to the one, two or three person small businesses which predominated at Bolsover Enterprise Park during this period. Nevertheless, on a wider canvas, in order to encourage private sector speculative development of business premises, potential rental revenues had to be worthwhile from the industrial property investors' point of view. As of 1990, little of this private sector speculative development was happening in the Bolsover area: the absence of such formed a part of the North Derbyshire Coalfield Partnership Project's case for public sector investment via the Rural Development Commission (North Derbyshire Coalfield Partnership, 1990).

The North Derbyshire Coalfield Partnership Project was the body formed by the local authorities in the area for the purpose of making a submission to the Rural Development Commission (RDC). This submission was aimed at securing public funds for a variety of projects ranging from industrial workspace units and environmental improvement schemes, through to social projects such as the partial funding of a community development worker to offer assistance to the relatively disadvantaged in society, such as the long term unemployed, and people in poverty.

The partners in the North Derbyshire Coalfield Partnership Project were Bolsover District Council, North East Derbyshire District Council, Derbyshire County Council, Derbyshire Rural Community Council, the Rural Development Commission (RDC) itself, British Coal and British Coal Enterprise. Submissions were made to the RDC by the Partnership Project in the 1990/91 and 1991/92 financial years, and parallel bids were submitted by Nottinghamshire

County Council and Rotherham Metropolitan Borough Council. The first approach to the RDC had been a joint one between Derbyshire and Nottinghamshire County Councils and Rotherham Metropolitan Borough Council in March 1989. That approach was aimed at highlighting the problems engendered by pit closures. These the partnership summarised as:

'Job losses, unemployment and poverty; poor environment and infrastructure; loss of individual and community confidence; little or no incentive to re-invest in the area' (North Derbyshire Coalfield Partnership, 1990).

Further south, a similar body to the North Derbyshire Coalfield Partnership Project was formed called the Leicestershire and South Derbyshire Coalfields Rural Partnership.

A second activity of Bolsover Enterprise Agency was to offer business counselling to site tenants, and potential site tenants. What should be stressed, however, was that this was within the managerial and knowledge limits of the Agency's limited number of staff. The Agency was the first to acknowledge that they were not managerial, business or marketing consultants. Their advice was often at the level of applied common sense to people embarking, or thinking of embarking, for the first time on a business venture. And, very often, the advice that would be given would be for the potential business person to consult someone else who did have a claim to managerial or technical competence.

Table 2

Businesses and paid employment on Bolsover Enterprise Park as of July 1991

Firm	Jobs (i)	Jobs (ii)	New or Tradition (iii)	Owner in coal in past (iv)	Why Park?	Business Before Park
1. Waste Disposal	2	0	Tradition	Miner	Local	Yes
2. Upholstery	2 (vi)	0	?	?	?	?
3. Design	2	2	New	No	Cheap	No
4. Weld	2	1	Tradition	No	Cheap	Yes
5. Sandwich	1	4	New	Coke	Cheap	No
6. Electronic	3	0	Tradition	No	Local	Yes
7. Coach A	3	0	New	Yes	Local	No
8. Coach B	2	1	New (vii)	Yes	Local	Yes
9. Food	3	0	MNC	N/A	?	Yes
10. Computer	5	0	New	No	Cheap Local	No
11. Accountant	1	1	?	?	?	?
12. Cables	7	3	New	No	Space	Yes
13. Fences	4	0	Tradition	No	Space	Yes
14. Drill Sharp	2	1	New	Yes	Coal	No
15. Pavement	1	0	?	?	?	?
16. Geologists	2	0	?	?	?	?
17. Graphic Design 2	4	0	Tradition	No	Cheap	Yes
18. Roofing	5	2	New	No	Cheap	No

Firm	Jobs (i)	Jobs (ii)	New or Tradition (iii)	Owner in coal in past (iv)	Why Park?	Business Before Park
19. Generator Repair	1	0	New	Yes	Local	No
20. Clothing 1	8	0	Tradition	No	Bought out firm	N/A
21. Bridalwear	1	0	New	Hus- band at pit	Local	No
22. Cater 1	1	0	New	Hus- band at pit	Local	No
23. Engineer	1	0	?	?	?	?
24. Predictive Engineers	40	0	New	Yes	Coal	No
25. Antiques	2	0	Tradition	No	Local	Yes
26. Leisurewear	10	0	Tradition	No	Cheap	Yes
27. Concreting	3	0	Tradition	No	Local	Yes
28. TV/Satellite Installation	2	0	?	?	?	?
29. Computer 2	2	0	Tradition	No	Local	Yes
30. Friction Material	5	0	Tradition	No	Cheap	Yes
31. Sportswear	1	2	Tradition	?	?	?
32. Scaffold	2	0	New	No	Cheap	No
33. Used Computer	5	0	New	No	Cheap	No
34. Silk Screen	1	0	New	No	Local	No
35. Sportswear 3	1	3	New	No	Cheap	Yes
36. Screen Printing	1	0	?	?	?	?
Total	138	20				

Notes to Table 2

(i)   full-time;

(ii)   part-time;

(iii)   'New' refers here to 'new entrepreneur'. This is
        defined here to mean a person not previously
        self-employed or running own business, but
        employed or unemployed elsewhere, and coming
        nearly to entrepreneurship in late 1980s or early
        1990s. 'Tradition' refers to someone who has an
        established career pattern of self-employment or
        running own business, for whom running a business
        in the late 1980s/early 1990s was not a new
        endeavour;

(iv)   A broad definition of 'coal industry' is applied
        here to include one owner/ manager whose previous
        employment was with the private sector Coalite
        company, producers of smokeless fuels and
        chemicals from coal;

(iv)   'Locals' here means the owner/manager chose the
        site because he/she/family were from the locality
        and wanted to operate from there;

(vi)   Estimate;

(vii)   Taking on already established business.

    Thirdly, Bolsover Enterprise Agency
effectively acted as a coordinator of
multi-agency activity in the immediate locality.
The Rural Development Commission, for example,
made itself available as a business advisor to
companies on the Park, with the Enterprise
Agency making sure that all the companies knew
of the RDC's existence. British Coal Enterprise,
the job creation arm of British Coal, was
another agency operating in the locality through
the mechanism of business advice and business
loans at that time at usually favourable rates
of interest. Bolsover Enterprise Agency
sometimes acted as the conduit to British Coal
Enterprise.
    The involvement of the Rural Development
Commission — which might provoke images of
horse-shoeing operations and thatched roofing —
in regenerating areas associated with such a
heavy industry as coal could engender some

105

surprise. The Rural Development Commission was Britain's oldest quango, dating from 1909, and was dedicated to the alleviation of social and economic problems in rural areas (Mawson, 1986). Its involvement here emphasised the fact that coal extraction was a rural, or at least semi-rural, industry. There were some notable exceptions to this. Howell, for example, noted that 'in Lancashire, mining was an industry of city and town as well as of pit villages' (Howell, 1989, p.6); and parts of the Nottinghamshire coalfield provide another exception, with the former pits of Babbington and Clifton, for example, within the boundaries of the City of Nottingham. The Durham, Northumberland, South Wales, Yorkshire and north Derbyshire coalfields, however, provided abundant evidence of the industry's (or former industry's) semi-rural status. Indeed, the fact that coal mining provided an underground location for industry traditionally meant that pressures for industrial and commercial development above ground were correspondingly less. This both helped to preserve the green and pleasant status of some parts of the coal-getting regions — though other parts were despoiled by waste tipping, opencasting and other environmental degradation — while, in some cases, it restricted the availability of land for post-coal mining industrial development. Green belt constraints on the use of land for industrial or commercial purposes, where these were originally introduced to prevent despoilation, were cited as problematic by authorities in South Yorkshire in a 1989 survey by Roberts and Green (1990), for example. The lack of an adequate transport infrastructure, because of the geographically isolated nature of many mining communities, was another factor held by local authorities to be prohibiting economic regeneration, particularly in the upper valleys of the South Wales coalfield and the coalfield communities near Wakefield in Yorkshire. Subsidence was a problem cited by Scottish local authorities in the 1989 survey as militating against industrial development. Other local authorities, particularly in the north west of England, cited

British Coal's slowness at relinquishing colliery and associated sites which potentially were some of the largest available sites for industrial development. A less important but still significant factor preventing industrial development was the absence of flat sites for such development (Roberts and Green, 1990).

Bolsover fitted well the Rural Development Commission's criteria for assistance. Its Annual Report for 1987/88 noted that:

> 'Our help is needed particularly in areas which have been dependent on one major industry or employer which are now in decline or shedding labour' (Rural Development Commission, 1988, p.17).

The same report noted that the RDC had given 'extra help to the Durham coalfield area' (Rural Development Commission, 1988, p.17).

The report also stated that the RDC's business advice 'services are available to firms employing not more than 20 skilled workers based in the countryside or country towns generally with a population of not more than 10,000' (Rural Development Commission, 1988, p.23).

A central function of the RDC's activities in the late 1980s and early 1990s was the attempt to stimulate small privately-owned businesses in small towns. The Annual Report 1987/88 emphasised this explicitly:

> 'Wherever possible, we seek to build on and stimulate indigenous enterprise and self-help. We encourage the private sector to operate both by creating the right framework and by operating in partnership with it' (Rural Development Commission, 1989 p.6).

This operating ethos chimed very well with the approach of both Bolsover Enterprise Agency and British Coal Enterprise, both of which sought regeneration through private sector small business stimulation.

The North Derbyshire Coalfield Partnership's regeneration strategy also reflected this ethos of private sector investment stimulation and an emphasis on small business promotion. According to them, given the resources available to local authorities the 'way forward .. [was] ... to stimulate private sector investment in the area', and, as part of a plea for funds from the RDC:

'To create more jobs additional premises are .. required to appeal both to new small businesses within the area and to attract new businesses from other areas' (North Derbyshire Coalfield Partnership, 1990).

The immediate question that arose was whether this was the most appropriate regeneration strategy for such a locality.

4.4 Bolsover Enterprise Park

All the 36 tenants on Bolsover Enterprise Park as of July 1991 were owner-managed small businesses, except for one distribution depot being run by a multinational.

Interviews of 28 owner/mangers of the total of 36 businesses on Bolsover Enterprise Park indicated that seven of these, or 25 per cent, had been employed in the coal industry at some stage in the past. Only three of these owner/managers had, however, come from non-supervisory, non-managerial or non-technical grades within the coal industry. In other words, for only three would the description 'miner' be anything near appropriate. The businesses of two of the seven were still heavily reliant on selling services to the deep-mined coal industry, and therefore reliant on its continued existence both locally and nationally for their own existence.

In a more balanced local economy, ie one not reliant — or not formerly reliant — on one industry, a figure of 25 per cent of owner/managers of small businesses emerging from the coal industry would appear encouraging, if one was seeking to foster small business creation as an alternative to the former basic industry. 28 interviews, however, was a (necessarily) small census, rendering the concept of '25 per cent' of them formerly employed in the coal industry somewhat less than significant. Further, given the nature of this local economy, and the hugely-dominant role played by coal in the immediate past, and given that the percentage of owner/managers dropped to 11 per cent, (still a less than significant statistic from a social science viewpoint, due to small census size), if managerial, supervisory and technical grades were excluded, the figures of Witt (1990) seem to be corroborated. Witt conducted a survey of miners made redundant in the Barnsley/Wakefield area, which was published in 1990. The evidence from that survey was that the setting up of small businesses, or self-employment, may not be the answer to the job needs of former coal industry employees. The study found between 7 and 10 per cent of redundant miners in self-employment just over a year after pit closures.

It is important to recognise, however, that most of these people had helped supervisory jobs at the pit and could not therefore be accurately characterised as miners. As Witt 1990 noted:

'Among the manual workers, only the craftsmen were well represented in self-employment, particularly electricians. For most non-craftsmen, self-employment does not appear to be a viable response to redundancy.'

By itself, this did not mean that attempting to stimulate the small business sector was the 'wrong' strategy. It may well have been, for example, that those involved in the regeneration efforts — the Enterprise Agency, British Coal Enterprise, the Rural Development Commission — believed that economic benefits would 'trickle down' to former coal industry employees. There was little evidence of this happening, however, in Bolsover. Alternatively, there may have been little else that the regeneration agencies could do **other** than attempt to stimulate the small business sector. Their resources were limited. They did not have the 'power' to attract — or force — a large scale employer to locate in the area, as central government might through a more-or-less coercive version of regional policy discussed earlier.

To some extent, the evidence from Bolsover supported the findings of Thomas (1989) on the effectiveness or otherwise of stimulating the small business sector as, in that case, a means of reviving economically the South Wales coalfield. Thomas found that business assistance from British Coal Enterprise, for example:

> '.. tends to be taken up, not by ex-miners or by the unemployed of mining communities, but by individuals who, in may cases, were found to have had previous businesses experience, and/or be fairly highly educated, and to be in relatively well-paid jobs prior to starting up a business.'

4.5 The 'new' entrepreneur

There **was** some evidence in Bolsover, however, of the emergence of the 'new entrepreneur', defined here as meaning a person not previously self-employed or running his or her own business, but coming newly to self-employment/small business ownership in the late 1980s/early 1990s. Sixteen, or 57 per cent, of the interviewed 28 fitted this category and were without, in the terms used by Thomas, 'previous business experience' (see Table 2). Again, the (necessarily) small size of the interviewed sample should be noted: it rendered any inferences drawn on the reasons for 'new entrepreneurship' less than conclusive.

What was important were the reasons **behind** 'new entrepreneurship.' There is a world of difference between, on the one hand, being forced into it as an attempt at a means of economic survival, because of redundancy or through an inability to get a job through high levels of unemployment/surplus labour and, on the other hand, a positive desire to become self-employed and/or run one's own small business. At least 10, or 36 per cent, of the 'new entrepreneurs' had faced personal economic circumstances — redundancy, inability to get a job in their preferred locality, inability to

110

get a job at all — which had led them to attempt to seek a livelihood via self-employment and/or small business ownership when, in other circumstances, they may not have chosen to do so. If the mark of an emerging 'enterprise culture' — where people's **preferred** option is to be self-employed and/or own a small business (Turner, 1990) — then its emergence in Bolsover was fairly muted. With only two — or 7 per cent — of the 28 owners/managers interviewed having backgrounds in the non-supervisory, non-managerial, non-technical side of the coal industry **and** qualifying of the designation 'new entrepreneur,' the emergence of the 'enterprise culture' in this geographical, economic and social sector as a viable force for regeneration was virtually non-existent.

In economic terms, of substantial interest to the population of a locality is the existence or otherwise of paid employment. The 138 full time and 20 part time (equivalent, say, to 10 full time jobs), that existed on the Enterprise Park as of July 1991 were, of course, worthwhile and of value to the local economy. However, in numerical terms, they were a pale shadow of the 300 jobs lost to the immediate locality with the closure of the NCB administrative headquarters on the same site in 1987, to say nothing of the jobs lost in nearby pits in the 1980s and 1990s. Moreover, and dangerously for the local economy, 42 — or 30 per cent — of the 138 full time jobs were reliant on providing services to the deep-mined coal industry and therefore, as noted earlier, on the latter's continued existence. This would serve to question the sustainability of what limited economic regeneration there had been at this time. This was particularly so given that all three remaining North Derbyshire collieries were announced for closure in October 1992.

Another important question that arose related to how many of the jobs on the Enterprise Park were actually 'new' to the local economy.

Thirteen, or 46 per cent, of the 28 owner/mangers interviewed indicated that their businesses had **already existed**, with similar numbers of jobs, elsewhere in the nearby locality prior to the establishment of the Enterprise Park. Sometimes the businesses had operated from the owner/mangers' homes; sometimes from other industrial premises. At the time the research was carried out, businesses employing 50 people fitted this category. A reduction of 50 from the 138 jobs 'created' on the Enterprise Park left a figure of 88 jobs 'new' to the local economy: hardly an encouraging figure given the scale of deindustrialization and job losses in the locality.

The news for the Rural Development Commission and British Coal Enterprise on the usefulness of their operations in relation to these very small, owner-managed businesses was mixed, but on the whole not encouraging. The provision of advice to small businesses was a central element of the Rural Development Commission's regeneration strategy and nine owner/managers reported that they had advice from the Rural Development Commission, but only 2 reported that the advice was useful. Another 2 reported that it was not useful at all, and a further 2 reported that the advice was only moderately, or partially, useful. The remaining three that had advice were not impressed by what they had received.

NCB Enterprise, the original title adopted by British Coal Enterprise, made it clear in its initial publicity documents that it too was looking to make an impact on local economies through the provision of business advice. It would be:

'Making available a full range of managerial skills on an ad hoc basis to assist in counselling and advice to (enterprise) agency clients' (NCB Enterprises, 1985).

Nevertheless, only four of the sample had had business advice or 'soft' loans from British Coal Enterprise. Of the four, only one reported that the help from British Coal Enterprise was useful, indeed terming the soft loan that the business had received 'essential'. The remaining 3 did not find the help useful at all (see Table 3).

Again it should be noted that the number of people interviewed represented, of necessity, a small census. The inferences to be drawn on the usefulness of both the RDC's and British Coal Enterprise's activities cannot therefore be conclusive. It may well have been that the particular businesses established at Bolsover were not the kind of businesses **anyway** that would have benefited from this kind of business advice; or it may have been that the personalities running the businesses were not receptive to being given advice from any quarter. The likelihood remained, however, that there was a mismatch between what the economic regeneration agencies were offering, and what the small business owners needed.

It should be noted, of course, that the provision of business advice and soft loans was only part of the activities of British Coal Enterprise. It was also involved in workspace provision; retraining; land use conversion; job search for redundant mineworkers. That the reception of small businesses towards advice and loans in Bolsover in 1991 was cool did not, therefore, mark a condemnation of the entire activities of BCE.

Similarly, the ambit of the Rural Development Commission's operations was wider than the provision of advice to small businesses. It was also involved in financing workspace provision, environmental improvement, social programmes and promoting tourism potential. On the first two categories, it invested £631,640 towards a programme costing £1,049,500 during the 1990/91 financial year in the north Derbyshire coalfield (Derbyshire County Council, 1991).

If the provision of business advice by regeneration agencies was less than successful, a brighter aspect of the economic regeneration strategy at Bolsover Enterprise Park was the cost to the public purse. Measured in terms of cost per job, the £32,000 per year coming from public monies of the £150,000 annual budget of the Enterprise Agency, divided by the 138 full-time jobs on the Enterprise Park, came to a very modest £232 per job per year. Even if this was reduced to a figure of 88 full-time jobs, in line with earlier comments, the cost-per-job per-year was still around only £364.

4.6 Conclusion

Bolsover district and what was the north Derbyshire coalfield passed through a period of rapid and fundamental economic change in a very short space of time in the 1980s and early 1990s. Central to this change was a major run down in the coal industry in a locality where the economy had been dominated by it.

The effects of the activities of the 'central agents of economic regeneration' — Bolsover Enterprise Agency and, on Bolsover Enterprise Park, British Coal Enterprise and the Rural Development Commission — were very modest in terms of the fostering of economic activity and jobs to replace those lost economic activity on Bolsover Enterprise Park. As for the usefulness of the provision of business advice by the Rural Development Commission, and soft loans and business advice by British Coal Enterprise, the findings will be disappointing for those two agencies. That disappointment would be tempered however, by the fact that the interviewed sample was of necessity small, and by the fact that both RDC and British Coal Enterprise engaged in a much wider range of economic regeneration activities than those examined here.

There **was** some evidence of 'new entrepreneurship' at Bolsover Enterprise Park: more than half of the interviewed sample were coming for the first time to self-employment and/or small business ownership. This would look positive from the viewpoint of those seeing an expanding small business and/or entrepreneurial sector, and the encouragement of people to join that sector, as being central to economic regeneration. The finding was more problematic than might at first appear, however, given that over half of the 'new entrepreneurs' were engaging in small business activity simply because other economic/employment opportunities were not available. Birch (1987) argued that many new small firms in the USA could be categorised as what he called the 'income

114

substitutor' type. Here, people who could not find work with conventional employers started businesses simply to provide themselves with an income. It is possible that a similar situation applied across national boundaries: specifically to north Derbyshire.

It could be argued, of course, that in the recessionary climate of the late 1980s and early 1990s — May 1991 saw the highest level of company liquidations for over twenty-one years (Peat Marwick McLintock, 1991) — the effects of these 'central agents of economic regeneration' **could only** be modest, and that no one should have looked for anything else. That might have been a valid point: it did not detract from the urgency of finding a strategy that would provide for economic regeneration in the coalfields even in the economic climate of the early 1990s. Clearly, the validity of an economic regeneration strategy which took as its major focus — as at Bolsover Enterprise Park — the encouragement of entrepreneurship and small business development was called into question by this evaluation.

Essentially, in the economic climate of the early 1990s, with national economic recession compounded at local level by rapid decline in the coal industry, the economic forces operating within a locality such as Bolsover and, indeed, in the rest of what was the north Derbyshire coalfield, were centrifugal. People would increasingly commute to nearby bigger towns to find paid employment, or would move to find work. This was acknowledged by the 1990 Derbyshire Structure Plan (Derbyshire County Council, 1990). Evidence of population decline was also provided by the census figures for 1971 and 1981 for Langwith: over the decade, the population fell by 12 per cent. Amongst the 25 - 34 age group the fall was 25 per cent (Coalfield Communities Campaign, 1986).

Such a strategy of individual mobility was one that was usually available only to the relatively young, fit and skilled. Looking at what happened to Britain's inner cities in the 1960s, 1970s, 1980s and early 1990s, where similar centrifugal economic forces were operating, that did not bode well for declining coal mining localities such as this one.

Table 3

Business on Bolsover Enterprise Park
Responses to Business Advice from RDC and BCE

Table Comprised as of July 1991

Firm	RDC Advice	Useful	BCE Advice/Help	Useful
1.	No	N/A	No	N/A
2.	Not known	–	–	–
3.	No	N/A	No	N/A
4.	Yes	No	No	N/A
5.	No	N/A	Advice	No
6.	No	N/A	No	N/A
7.	Yes	50% Useful	No	N/A
8.	No	N/A	No	N/A
9.	Not known	–	–	–
10.	Yes	Yes	No	N/A
11.	Not known	–	–	–
12.	Yes	No	Advice	No
13.	Yes	'moderate'	No	N/A
14.	Not known	–	–	–
15.	Not known	–	–	–
16.	Not known	–	–	–
17.	Yes	Yes	No	N/A

Firm	RDC Advice	Useful	BCE Advice/Help	Useful
18.	Not known	-	-	-
19.	Yes	No	No	N/A
20.	Offered not taken	N/A	No	N/A
21.	No	N/A	No	N/A
22.	No	N/A	No	N/A
23.	Not known	-	-	-
24.	No	N/A	Loan	essential
25.	No	N/A	No	N/A
26.	Yes	No	Advice	No
27.	Not known	-	-	-
28.	Not known	-	-	-
29.	No	N/A	No	N/A
30.	Offered not taken	N/A	No	N/A
31.	Not known	-	-	-
32.	Yes	No	No	N/A
33.	No	N/A	No	N/A
34.	Not known	-	-	-
35.	Offered not taken	N/A	No	N/A
36.	Not known	-	-	-
TOTAL	9 YES		4 YES	

5 British Coal Enterprise: Bringing the 'enterprise culture' to a deindustrialized local economy?

5.1 Introduction

British Coal Enterprise (BCE), the job creation arm of British Coal, started life in October 1984. It followed the model established by British Steel Corporation (Industry), which was set up in 1975 under the then Labour government. BCE has had three major prongs to its job creation strategy: the provision of loans to companies at interest rates lower than banks; retraining of redundant mineworkers; provision of managed workshops as premises for small businesses.

This study analyses BCE activities in one particular locality: Carcroft, near Doncaster. Analysis of its activities on a nationwide basis have taken place elsewhere (Owen, 1988). The justification for this study of the impact of BCE on a local economy is that national-level analyses are, of necessity, broad brush. The picture they paint has to be one of generality: that x thousand jobs have been created at a cost of *x* thousand pounds each. Central to the approach here is that a real understanding of BCE's activities can be gained only by looking at its effects on a particular local economy: it is in these localities where the benefits or otherwise will really be felt.

BCE has already been criticised for operating over too wide a region, and providing assistance to companies in areas where there was little justification in terms of coal mine closures. Owen (1988) noted, for example, that assistance has been given to companies in 'a small part of Glasgow, Workington/Whitehaven, the Kent coalfield and two small areas of North Wales' where 'pit closures in recent years have had little effect on regional unemployment rates.'

Hudson and Sadler (1992) have examined elsewhere reindustrialisation policies pursued in Derwentside in the 1980s. That study is germane to this because a central point there was that the regeneration effort was attempting to replace jobs lost in the traditional industrial sectors with new jobs 'born of a flourishing enterprise culture.' This study, however, focuses on BCE's activities in a locality heavily and traditionally associated with coal mining: Carcroft. What was the nature of BCE operations there? Was the approach to job creation appropriate? Was adequate monitoring and evaluation of its own operations undertaken by BCE? They also noted that if regeneration policies were to be pursued successfully in the future, it was important to assess accurately changes in the local economy promoted by regeneration agencies. In that sense, reindustrialisation strategies pursued at the local level have had a national significance, because Britain in the 1980s and early 1990s had many deindustrialized localities. A successful regeneration strategy in one could be replicated elsewhere; policies for regeneration that were less than successful in one locality could be avoided elsewhere.

This work is based upon an interview with a senior official, at the local level, of BCE; and on telephone interviews with owners, managers, or owner/managers of 25 companies located on

Carcroft Enterprise Park. All the interviews took place in February 1991.

5.2 Coal: the scale of deindustrialization

The justification for public sector involvement in job creation was that the scale of deindustrialization in the coalfields has been vast, and was effected rapidly, in the late 1980s and early 1990s.

Moreover, as is well known, in many areas affected by pit closures, the coal industry was the mainstay of the local economy. Unemployment was, in any case, higher in some of these localities than elsewhere: one study estimated unemployment as being 30 per cent higher in coalfield areas that non-coalfield areas in 1988 (Owen, 1988, p.5). Pit closures were making a bad unemployment situation worse; their impact was damaging a large number of small, local economies.

5.3 The locality

Carcroft is a community of 4,000 people about five miles to the north of Doncaster. It is not a stranger to contraction in its basic industry: its own pit, Bullcroft, closed in 1970 with the loss of 690 jobs (*Guide to the Coalfields*, 1986). The coal industry nevertheless remained important to its economy. In Carcroft itself there remained, until after the 1984/85 coal industry dispute, central stores, workshops and a repairs depot, employing nearly 400 people, servicing other local collieries and maintaining houses owned by the then National Coal Board.

Elsewhere in the nearby locality there were pit closures in the 1980s and early 1990s. Brodsworth colliery, in 1984 the second biggest pit in the entire Yorkshire coalfield, employing 2,069 (Glyn, undated), and which closed in September 1990, was little more than a mile away. Bentley colliery, still in operation as of December 1991 and which saw a rundown in its workforce from 1,010 people in 1985 (*Guide to the Coalfields,* 1986) to just over 600 in June 1990, was two miles away. Three miles to the north, Askern colliery saw a rundown in its workforce from 1,262 (*Guide to the Coalfields*, 1986) in 1985 to around 600 in February 1991. Its total closure was announced in November 1991 (*Doncaster Star* 23 November 1991).

This rundown described was very much in the immediate vicinity of Carcroft. Travel a further five or six miles and the extent of the 'pit closure zone' became ever more evident: South Kirkby, Barnburgh, Hickleton Main, Highgate, Cadeby, Wath Main, Manvers Main, Yorkshire Main were all pits closed in the mid-1980s or the early 1990s.

Clearly, then, British Coal Enterprise was targeting its operations in the right place. What was also evident was the mammoth scale of the task facing it and other job creation agencies in this deindustrialized local economy.

5.4 BCE operations at Carcroft

The centre point of BCE's operations at Carcroft was Carcroft Enterprise Park. It was on the site of the central stores, workshop and repairs depot referred to earlier. All the property and the land was in the ownership of BCE.

Carcroft Enterprise Park had been in existence since late 1985/early 1986. It was the first of its kind operated by BCE, but British Coal's job creation agency must have been keen on the idea as they followed it by setting up Bersham Enterprise Centre in North Wales and Haig Enterprise Park in Cumbria. A similar operation was also put into place on the site of Ackton Hall colliery, in Featherstone, West Yorkshire, which closed in 1985; and on a site adjacent to Dinnington colliery, near Rotherham, scheduled for closure in early 1992 (*Coal News*, 1991). The popularity with BCE of this job creation response represented another justification for examining its effectiveness at Carcroft.

Activities at Carcroft Enterprise Park could be divided into three. Firstly, it was a straightforward industrial estate, differing from some other industrial estates in that, at the start of its operations, rents were apparently cheap, and BCE had 'recycled' some former NCB/British Coal buildings for successor commercial/industrial uses. Some of the smaller properties on the Enterprise Park were 'licensed' from BCE rather than leased; businesses licensing properties in this way needed to give only one month's notice to quit, and therefore did not have large sums of capital tied up in a lease. As of February 1991, there were 44 companies on the Enterprise Park employing at maximum around 270 people. Twenty eight of these companies were in the straightforward industrial estate part of the Enterprise Park, employing 238 people. Sixteen businesses were in the Acorn workshops discussed below, employing a maximum of thirty two.

Secondly, BCE engaged in the retraining of former British Coal employees at Carcroft Enterprise Park. Mainly, this was retraining in welding and electrical skills. No figures were available for the number of people who have been retrained in this way at Carcroft. BCE would not divulge the number.

Thirdly, BCE provided managed workshop space for small businesses. These managed workshops were called 'Acorn' by BCE, for obvious reasons.

As of February 1991, there were 22 managed workspace units in 10,000 square feet of space. A manager was attached to Acorn to deal with the administration, payment of heating and lighting bills and so forth; a receptionist was similarly attached, and each unit had telephone facilities.

The idea was that those wishing to set up in business, perhaps for the first time, could move into the Acorn workspaces without the burden of having to find large capital sums to finance property. Additionally, they could leave with one month's notice, so again, large capital sums were not tied up in property ownership or leasing. The objective was to ensure maximum flexibility so that, optimistically, when companies were ready to move on to bigger premises, it was easy for them to do so. Pessimistically, if the businesses folded, they did not lose large sums of money tied up in property.

The philosophy which underpinned the economic regeneration strategy represented by Acorn was that the size of the small business sector could be increased. As small firms grew bigger, this would form the basis for economic growth in the locality.

5.5 The enterprise culture

In its emphasis on stimulating the small business sector, the regeneration strategy pursued at Carcroft neatly linked in to the 'enterprise culture' policy thrust pursued and advocated by the Conservative central government, particularly when led by Mrs Thatcher. The concept of an 'enterprise culture' has been defined, in a broad sense, as the elevation of the status of private sector organisations to models for public sector organisations to emulate; the removal of government regulations on business and the concomitant extension of the free market; the redefining of the objectives of the state education system to emphasise the vocational and the promotion of 'self-reliant' individuals (Kent, 1991; Morris, 1991). For our purposes, in relation to the regeneration of a local economy, the definition can be narrower: it can be reduced to a general encouragement of people to set up their own businesses (Turner, 1990).

A major aspect of the 'enterprise culture' that the Conservatives were hoping to create was an economic environment within which the norm was to set up one's own business rather than looking to other already existing commercial organisations to provide paid employment.

The creation of a local 'enterprise culture' at Carcroft was emphasised as being an objective of regeneration policy by the BCE official interviewed for this work.

Previously, this locality had been dependent upon one major employer, and economically it was devastated when rapid contraction hit the coal industry. The objectives of BCE's regeneration philosophy were to provide a broader-based local economy, and to promote that through the encouragement of a more robust small business sector.

Before further analysis is attempted, there is a need to address the issue of whether or not this was the correct regeneration strategy for a local economy such as this. To do so it is important to assess the nature of the local economic base upon which the attempt is being made to superimpose an 'enterprise culture.'

Certain facts were well-known and needed little rehearsal. First of all, the skills associated with coal mining were specialised and not easily transferable to activities outside the industry. As noted by Meegan (1991), in his analysis of some of the attempts to regenerate a non-coal mining area but one which has had its own severe problems of deindustrialization, Merseyside, the rationale of trying to create an 'enterprise culture' in a local economy with a circumscribed skill base and a depressed local income and demand was called into question. A redundant plumber may well be able to set up his own central heating installation business, for example, if the local economy was otherwise strong and could sustain it. Such transfer of skills was more unlikely for a man who was used to hewing coal since his working life began; especially if the local economy was depressed.

Secondly, a substantial number of mineworkers and former mineworkers had only a fairly limited formal education. This was not to blame mineworkers for this, as it could be argued, for example, that the schools system which they went through in mining localities was geared almost solely to providing future mineworkers (Allen, 1981, p.75). Many mineworkers would claim that this was the case. Nevertheless, in the absence of a large scale and intensive retraining programme, a huge swathe of potential entry opportunities into the elusive 'enterprise culture' were cut off by this factor as it limited the vocational activities that might be entered into.

There was a third, connected, problem with trying to establish an 'enterprise culture' in a local economy that had been so heavily and traditionally associated with coal mining. It could be cogently argued that once workers were involved in coal mining for a number of years, many began to believe, rightly or wrongly, that they were capable of nothing else. This was noted in a study of coal mining closures affecting the Shotts locality in Lanarkshire as far back as 1949 (Heughan, 1953, p.120), for example, and was attested to by people during the research for this study. A survey in 1989 of miners made redundant by the closure of the Woolley and South Kirkby collieries in the Barnsley/Wakefield area in 1988, noted that the largest group of economically active former mineworkers — a third — had returned to mining by working for private sector contractors to British Coal. The same study noted that 'the number of former miners establishing small businesses after redundancy is low' (Witt, 1990). The apparent belief by some mineworkers and former mineworks that coal mining was the only occupational activity they could pursue, whether reflecting psychological factors or actuality, could not but militate against the objective of the creation of a local 'enterprise culture.'

In practice, understandably, no discrimination relating to former occupation was applied by BCE as to who occupied the Acorn workspaces, or any other commercial property at Carcroft Enterprise Park. There was no requirement to have been previously employed by British Coal. The 'pioneers' of the enterprise culture then — people setting up small businesses — could have had a career history outside coal mining. But the fact remained that it was former mineworkers, and those who would have expected to enter the coal mining industry but never had the chance, who needed paid employment.

Elsewhere in the Yorkshire coalfield, evidence available would lead to scepticism that an 'enterprise culture' was in fact being fostered in the wake of deindustrialization. The level of take-up of Enterprise Allowance Scheme grants — which provided small weekly grants for the unemployed, or those facing unemployment, to start up small businesses (Turner, 1988) — could be taken as one indication of the emergence or non-emergence of an 'enterprise culture,' for example. But in four of the 'Five Towns' of Wakefield, Pontefract, Normanton,

126

Castleford and Knottingley, take up of
Enterprise Allowance Scheme in 1985 was 30 per
cent below the rate expected by Wakefield Job
Centre officials; in Hemsworth/South Elmsall it
was 60 per cent below (see Table 4) (The Centre
for Employment Initiatives, 1986). All these
localities have been heavily associated with
coal mining in the past.

Table 4

Actual and expected take-up of Enterprise Allowance Scheme 1985 in the 'Five Towns'

Job Centre	Actual		Expected
Wakefield	127		110
Pontefract	40)		
Normanton	11)		
Castleford	20)	79	
Knottingley	8)		
Hemsworth	9)		62
South Elmsall	18)	27	
TOTAL	233		

Source: The Centre for Employment Initiatives (1986);
          original source: Wakefield Job Centre

5.6 BCE and evaluation

BCE's Annual Review for 1989/90 claimed BCE had helped to create 'new employment opportunities' exceeding 61,000 since its operations began (British Coal Enterprise, 1990a, p.2). BCE has been criticised, however, for, at best, being over optimistic, and at worst boosting, the number of new jobs that it claimed responsibility for having helped to create. Hudson and Sadler (1987), in one of their studies of reindustrialisation measures in former coal and steel areas, for example, argued:

'... it is not clear how many of the "new jobs" created are still in existence; how many are part-time or full-time; how may have been supported by both BC(E) and BSC(I) and other agencies, and are thus double-counted in claimed employment totals; how many pay a similar wage to that which they supposedly replace; and how many have been taken by ex-coalminers and steelworkers.'

One of the central findings of this research was that at Carcroft Enterprise Park no systematic evaluation was carried out of, for example, the fate of companies that had moved out of the Acorn workshops into premises away from the Enterprise Park. In the vast majority of cases, contact between BCE and companies formerly located in its premises was lost completely. Yet each job provided in the past by companies located at Carcroft was assumed to be still in existence, whether or not it was in fact, for the purpose of counting BCE's 'new employment opportunities.' It was on the basis of figures collected from local operations such as the one at Carcroft, that BCE aggregated and provided the figure of 61,000 'new employment opportunities' as of 1989/90. Given that some companies which started off on BCE premises, or were in some other way connected to BCE, would have expanded employment levels, and others would have contracted or ceased trading whatsoever, 61,000 as a job creation figure must be considerably wide of the true number.

As noted by Storey (1990), the 1980s and early 1990s saw a dearth of evaluation of the effectiveness of job creation agencies such as this. And, as he further pointed out, good quality evaluation was essential if policy-makers were to know which strategies provided the highest number, the best quality and the most cost-effective jobs. There was, after all, large sums of public money invested in job creation schemes such as the one operated by BCE. In its 1989/90 Annual Review, for example, BCE noted that it had committed funds in excess of £60m to job creation projects to that date.

129

Providing an analysis of cost-per-job for projects such as the Carcroft Enterprise Park was not an easy task for an outside researcher, however. The extent to which BCE believed in open government, at least in relation to its own operations, appeared to be limited. Firstly, BCE did not and was not willing to break down its aggregate budget into its separate allocations to individual projects, such as Carcroft Enterprise Park. Secondly, it did not and would not reveal which companies received what sums in the way of low interest rate loans. Understandably, companies themselves which have received such funds were often unwilling to reveal any information about them. Therefore the outside researcher reaches an impasse when it comes to the assessment of cost-per-job.

5.7 The outcome

Some points could be made, nevertheless, on the outcome of BCE activity at Carcroft.

As of February 1991, Carcroft Enterprise Park had a total of 28 businesses on its straightforward industrial estate part, providing about 238 full time jobs or full time equivalent jobs. In addition, there were 16 one or two person businesses in the Acorn workshops, providing a maximum of 32 jobs. The employment total at Carcroft Enterprise Park was thus of the order of 270. Businesses there ranged from reconditioning lorry hydraulic systems; engineering, joinery and welding services; the manufacture of diesel generators; the manufacture of chemicals for the building and civil engineering industry; through to the manufacture and repair of narrow boats. Most of the businesses were in 'traditional' industrial sectors or involved in the servicing of 'traditional' industry. Only three or four companies, employing very small numbers, could lay any claim to being at the high technology end of industry, and these were mainly in activities such as computer repair.

Of course, not all of the 270 jobs at Carcroft were 'new' jobs, attributable to the activities of BCE. Evidence from there suggests both considerable 'displacement' and 'deadweight'. The former referred 'to the situation that jobs may be created in one place at the expense of jobs in another place.' The latter 'occurs when the assistance given by the public body does not affect the behaviour of the project which receives the assistance: the investment would have been the same even without the assistance' (Coulson, 1990).

Both displacement and deadweight were to be expected in most 'job creation' schemes. In a territorially-specific scheme, firms would be motivated to move from one place to another to pursue their own advantage — say in securing better or cheaper premises — and in that indirect way, by boosting the prospects for the firm concerned, the prospects for the local economy might improve at least modestly even if 'displacement' occurred. This might not be the case, of course, if the improvement in the prospects of that one company damaged the market position of a competitor in the nearby vicinity. In that case the marginal improvement to the local economy of one company switching premises might be cancelled out as the competitor shed jobs. Of the 25 firms contacted, at least 11 would fit into the category of displacement or deadweight: 6 were local relocations; four were new branches of national or local firms wishing to expand in this locality **regardless** of any governmental or quasi-governmental activity; one was a long-established business which had gone bankrupt and had been restarted with a new name, making the same products in exactly the same place (see Table 5).

By far the largest category of the 25 companies contracted, however, classed themselves as 'new businesses.' Fourteen of the 25 companies, or 56 per cent, counted themselves as 'new businesses' that had not been trading anywhere else prior to Carcroft Enterprise Park.

The total number of jobs associated with a regeneration project could be referred to as 'accommodated' jobs: this involved a count of the total number of jobs, including deadweight and displacement. Jobs which would not have existed in the absence of governmental or quasi-governmental regeneration activities were referred to as 'attributable' jobs (Martin,

Table 5

Categorisation of 25 firms at Carcroft
Enterprise Park
as of February 1991

Local Re-location	New Branch	New Name	New Business
6	4	1	14
24%	16%	4%	56%

1989). Where there was employment growth in a company transferring from elsewhere, the jobs associated with that growth could be classed as 'attributable.'

Amongst the 28 businesses on the straightforward industrial state part of Carcroft Enterprise Park, there were 238 'accommodated' jobs, as noted earlier. One hundred and forty two of these could be classed as 'attributable' jobs on the above definition. An impressive number of the attributable jobs were associated with 'new businesses': 97 of the 142, amounting to 68 per cent. Indeed, 97 related to the 238 'accommodated' jobs is still an impressive 41 per cent of total jobs.

Whether or not this level of new business creation was consistent with the creation of a local 'enterprise culture' depended on why people were choosing to set up businesses. If they were choosing to establish businesses in preference to other options — such as paid employment in an already existing public or private sector organisation — that might be a sign of an emerging local 'enterprise culture' If, on the other hand, the establishment of a business was a reflection of a lack of other alternatives, because unemployment was high in a local economy, then the emergence of a thriving 'enterprise culture' may not have been indicated. In the absence of other, alternative jobs, the setting up of a business could reflect a desire to survive financially, rather than being an expression of previously pent up entrepreneurial motivation.

The evidence from Carcroft was not conclusive either way. Three of the 14 people starting new businesses had been made redundant from the coal industry. Establishing a business was clearly an attempt at economic survival. Three others declined to give a reason or did not appear to have a reason. Two wanted to compete in the same product/service sectors with former employers. One saw self-employment as more satisfying than working for someone else. Another owner-manager had a personal tradition of being in business, and his activities at Carcroft Enterprise Park were another expression of that. One owner/manager said that being a small businessman 'meant advancement'; another claimed his motivation stemmed from being a 'follower of Thatcher.'

5.8 Conclusion

Some form of activity aimed at job creation was vital to local economies that had been heavily dependent on coal mining and had suffered pit closures. It was clear from observing somewhere like Carcroft that such localities, with their often tightly circumscribed skills base and limited local demand for goods and services, and distance from areas of economic buoyancy, were not going to recover by themselves, as might have been imagined by a free market economist.

There were certain positive aspects to BCE's activities at Carcroft. Firstly, land vacated by the coal industry was not left derelict with the attendant damaging consequences that could have had on the physical environment. Secondly, at least some attempt was made by BCE, within limited resources, to stimulate economic activity at Carcroft. In the absence of this, the situation at Carcroft would have been unlikely to lead to 'spontaneous' economic regeneration, and unlikely to lead to a large influx of inward investment. The conditions were against it: it was a small community; off the beaten track; formerly heavily dependent upon heavy, traditional industry, where the skills of workers were not easily transferable to other industrial sectors.

The activities of an organisation like BCE, however, unless its budget and activities were substantially increased, could constitute only a partial and marginal response to a situation of rapid, large scale deindustrialization. In other words, the 270 jobs 'accommodated' at Carcroft Enterprise Park as of February 1991 were pretty small beer to the jobs lost in the coal industry in the 1980s and early 1990s.

A weaker aspect of BCE's activities was the absence of coherent evaluation of its job creation strategies. A tighter evaluation procedure need not have been expensive, and it would have guided BCE towards more efficient job creation activities. In addition, BCE appeared to be overly secretive in relation to some of its activities: on the budget accorded to particular Enterprise Parks, for example, or the number of people retrained in welding or electrical skills. This prevented an outside researcher from carrying out a comprehensive evaluation.

A further conclusion was that a large question mark remained over the 'enterprise culture' regeneration strategy when applied to localities such as this. There was no doubt that there were people new to business at Carcroft. Even if this was a sign of an emerging 'enterprise culture' in the locality, however, it was unlikely that would become so robust as to be sufficient to revive the local economy. An enterprise culture might be more likely to thrive in a buoyant local economy with a small business tradition, rather than a newly deindustrialized local economy where workers had been traditionally sellers of their labour to others. In those circumstances, attempting to foster an 'enterprise culture' might usefully form an adjunct to a more direct economic regeneration scheme, rather than being a central part of it. The crux of the problem was that deindustrialization was visited upon large numbers of communities like Carcroft without there being in place an adequate mechanism to deal with its social and economic consequences. BCE itself was an after thought to deindustrialization, being established eight months into the 1984/85 coal industry dispute. Pit closures figured prominently on the political agenda throughout the 1980s; what the consequences were and have been for pit town communities, until October 1992, clearly did not.

6   An enterprise zone in a pit closure zone: The politics of industrial subsidies

6.1   A 'politics' of industrial subsidies?

The juxtaposition of an enterprise zone scheme in a pit-closure zone allowed for a comparative analysis of two potentially competing job creation/maintenance policies.

The location reviewed is the coal mining town of South Kirkby in West Yorkshire. It had, on the one hand, a Thatcher government-initiated and sanctioned enterprise zone in operation since August 1981. And also had, on the other hand, the South Kirkby-Riddings coal mine complex, which was closed on the grounds that it was 'uneconomic' in March 1988.

Both operations had factors in common. Both made claims on central government for financial assistance. They were adjacent to each other, and served the same community. Both provided paid employment in an area that was relatively economically deprived even before further pit closures affected the area in the mid and late 1980s.

The exercise in comparative analysis detailed here is essentially a study of policy 'outcome'; in other words, a study of what actually happened as a consequence of policies pursued. This can be, but need not necessarily be, the same as what a government actually intended to happen (Heclo, 1972). The enterprise zone policy in Britain, for example, was, after all, an experiment, albeit one into which large amounts of both financial and political capital had been invested by the Conservative government of Mrs Thatcher.

The intention of this chapter, then, is to address the question of whether or not the enterprise zone in South Kirkby 'worked'. Did it provide new jobs for people in a pit-closure zone, and if so, at what cost? In particular, how did the cost-per-job to the taxpayer for the enterprise zone compare to the cost-per-job involved in subsidising the local coal industry? Were there grounds for arguing that some claims for financial assistance from industry/business were met, while others were refused, because some claims were more **ideologically acceptable** than others to the Conservative Government led by Mrs Thatcher, rather than on the grounds of objective economics or public expenditure criteria? In short, was there a 'politics' of industrial subsidies in existence in the 1980s and early 1990s?

6.2 The location

South Kirkby was a free-standing 'pit village' of about 10,800 people in West Yorkshire as of 1990. It was under the administrative control of the City of Wakefield Metropolitan District Council, but was actually geographically closer to Barnsley in South Yorkshire.

South Kirkby's links with coal mining go back a long way. The shaft at the South Kirkby pit was sunk in 1880. To the east, west, north and south of this town was traditionally coal mining territory. The economy of the locality has been overwhelmingly dependent on coal for as long as anybody can remember.

South Kirkby came within the parliamentary

constituency of Hemsworth which, as late as 1985, could still boast six pits within its boundaries: Acton Hall, Kinsley Drift, Nostell, South Kirkby, Ferrymoor-Riddings and Frickley-South Elmsall. As of 1992, only Frickley-South Elmsall remained. Frickley was one of the 31 pits announced for closure by the Government on 13 October 1992 (see *The Guardian*, 14 October 1992). Yet as late as August 1990, British Coal had seen fit to describe Frickley as a 'big hitting colliery ... there is no talk of it shutting ... the pit will last well into the next century' (British Coal, 1990).

In 1984, more than 50 per cent of the jobs in the locality were in coal mining (CWMDC, circa 1988/89). The rapid pit closure programme since then clearly emphasised the need for a boost to this locality's economy in order to reduce unemployment and increase prosperity.

Wakefield Metropolitan District Council estimated unemployment in the locality as standing at 24 per cent in March 1988, for example (*Hemsworth and South Elmsall Express*, 20 March 1988). Even prior to the pit closures of the mid and late 1980s, the locality was an economic black spot: unemployment was estimated at 22.8 per cent for South Kirkby and the surrounding area in July 1982, compared with a national unemployment figure of 12.9 per cent (*Hemsworth and South Elmsall Express*, 29 July 1982).

So it was here that the enterprise zone policy came in. The area needed jobs, and it needed a policy mechanism that would introduce **structural** changes in the nature of the industry in the locality. It needed diversification. For far too long the area had been dependent upon a single industry, moreover an industry in the 1980s which had entered a terminal decline in many localities.

Wakefield council's application to central government for enterprise zone status to be granted to Langthwaite Grange Industrial Estate at South Kirkby was not, however, made because they anticipated pit closures. One reason for the application was that Wakefield Metropolitan District was to lose Intermediate Area Status, under which companies could at the time apply for financial assistance to central government under national regional policy for locating or expanding in the area, in July 1982. In fact, Wakefield council were bullish about coal's prospects, and what they were really fearing was labour-shedding as a result of mechanisation in the mines:

'Whilst the renewed importance of coal as an energy source has ensured the future of the mining industry, the continuing improvements in productivity in the industry mean that employment in mining continues to fall' (CWMDC, September 1983).

What was to happen in actuality was that the region as a whole was to become an economically-blighted 'pit closure' zone. After the year-long coal industry dispute over pit closures had ended in March 1985, between then and October 1990, 7 pits within a five mile radius of South Kirkby were closed: Barnburgh Main, in 1989; Brodsworth, in 1990; Darfield Main, in 1989; Hickleton Main, in 1988; Kinsley Drift, in 1986; South Kirkby Riddings, in 1988; Wath Main, in 1987.

In the absence of any other economic initiative, the enterprise zone created at South Kirkby remained, as of 1990, the major attempt to regenerate the economy of that immediate locality. And, to reiterate, it was a central government-approved attempt at regeneration, involving financial subsidies to companies from central government. By way of a contrast, the extent to which the Conservative government of the time was prepared to invest financial and political capital intended to ensure pit closures, including fighting the year-long battle against the NUM, should be noted (Goodman, 1985).

6.3 The enterprise zone idea

It was in 1979 that Peter Hall, then Professor of Geography at Reading University, first canvassed the idea of creating 'mini-Hong Kongs' in Britain as a means of reviving the economies of inner-urban localities. In a speech to the British Royal Town Planning Institute, Hall argued that inner-urban decline in some British cities had become so advanced that orthodox approaches to regeneration would be insufficient (Hall, 1982; Lawless, 1989). What was needed was a radical approach commensurate with the severity of the urban problems in some areas.

Hall advocated a limited number of what he termed 'free ports' which could be established in inner urban areas.

His was a radical vision of deregulation from government: these mini-Hong Kongs would emulate the free enterprise economy of the Crown Colony; there would be few if any restrictions on what businesses could do, in terms of wages they paid, the goods or services they traded in, the way they developed the land.

When Hall first put forward his ideas, Sir Geoffrey Howe, then Shadow Chancellor, took them up with some enthusiasm. The plan fitted in well with the Conservative Party's vision of a thrusting, dynamic, free enterprise economy liberated from the 'dead hand' of state interference. And, on taking power, an attempt was made to put a version of the idea into practice. Initially, eleven such 'enterprise zones', as they were called, were created in 1981 under provisions allowed for in the 1980 Local Government Planning and Land Act. A second round of fourteen further enterprise zones was announced in December 1982; by 1988, 26 were operating in Britain.

These zones, however, were far removed from the mini-Hong Kongs Hall had advocated. Howe's plan did not do away with the welfare state in the zones, nor were health and safety requirements reduced or the provisions of the Employment Protection Act removed from firms in these zones.

Instead, the enterprise zones were limited to a ten year existence, and provided the following benefits — effectively industrial subsidies — for that time: exemption from local authority rates on industrial and commercial property; 100 per cent capital allowances in relation to corporation or income tax for capital expenditure on the construction, extension or improvements of industrial and/or commercial buildings. A further major benefit, available only in enterprise zones until the total abolition of the tax in 1985/6, was exemption from development land tax (Bennet, 1990). There were also some more minor benefits: simplified planning procedures with concessions on the development that would be allowed; fewer government forms to fill in; certain customs facilities processed as a matter of priority; exemption from industrial training levies and from the need to supply information to industrial training boards, when these existed.

Hall (1982) argued, however, that even the more limited version of enterprise zone implemented by the Thatcher government would have beneficial results. The following two quotations from Hall (1982) summed up his expectations for them:

> 'The real argument for enterprise zones is that they will encourage enterprise ...'

> 'Starting perhaps with relatively low-level developments involving little technical skill, they would progressively sophisticate to become independent centres of technological innovation.'

Above all, if the economic 'threat' from newly industrialising countries was to be countered, there was a need for an industrial policy in Britain that would:

> '...involve ... an industrial policy based not on support of declining industries, but an encouragement of industrial innovation.'

So, in that sense, the idea of having an enterprise zone in a pit closure zone fitted Hall's thesis perfectly: it would be a source of enterprise and innovation in an area of traditional, declining, industry.

Further, South Kirkby fitted Hall's thesis in that the skills of the people there related to coal mining, the traditional industry, but might be inappropriate to some other form of

141

economic activity:

> 'Their skills (or lack of skills) were
> readily useable in an early industrial
> economy, but are no more. If we try to
> force them to catch up, we may do them —
> and us — no good, and perhaps a great deal
> of harm' (Hall, 1982)

Hall believed that his enterprise zone proposals would enable people with skills useable in an earlier industrial economy to use those skills in a present industrial economy.

In another sense, South Kirkby did not fit Hall's thesis at all well. South Kirkby was not an inner city area by any stretch of the imagination. It may have shared, in some ways, some of the problems of the inner-urban areas — unemployment and relative poverty, for example — but in the other ways it was completely different. The housing stock was reasonable, for instance, and was comprised of **houses** rather than multi-storey flats. The population was predominantly white and racially homogenous, unlike many British inner city locations. Rather than inner urban, it would have been more accurate to describe the locality as a collection of small towns and pit villages, (of which South Kirkby was one), surrounded by farmland and wooded countryside. Because coal mining meant effectively that a locality's industry was underground, it allowed in many areas for the maintenance of a green and pleasant environment, qualified of course by the development of spoil heaps.

6.4 Reality and analysis

Enterprise zone status was designated to an existing industrial estate at South Kirkby — Langthwaite Grange Industrial Estate — from August 1981. Enterprise zone status was extended in 1983 to two further, smaller industrial estates a few miles away, at Kinsley and Upton. This study concentrated on Langthwaite Grange, however, because: Langthwaite Grange was the largest of the three enterprise zone sites, at 140 acres, compared to

83.5 acres for the other two sites combined; it was the original site; it allowed for a direct comparison to be made on a cost per job basis with the South Kirkby Riddings pit complex, which was adjacent.

Four groups of companies and individuals gained from benefits available in the enterprise zones: occupiers of property; investors in property and/or business; property developers; landowners (PA Cambridge Economic Consultants, 1987). No attempt is made to differentiate between them here: all benefits are regarded as subsidies to private sector business.

An investigation into industrial subsidies paid to companies on an enterprise zone became, in effect, a work of detection. 'Discovering' subsidies paid to companies was not straightforward. The Inland Revenue, for example, was prevented from revealing the details of tax allowances received by companies to third parties by Section 182 of the Finance Act 1989. Companies themselves were not, for the most part, forthcoming about tax allowances they had received or other forms of subsidy.

This investigation, then, proceeded by another route. The rateable value of properties **was** public information, and was kept by local authorities, in this case the City of Wakefield Metropolitan District Council. The rateable values remained unchanged since 1979, business properties being revalued for rateable purposes in 1990 as a result of the introduction of the National Business Rate. In Wakefield, it was estimated by the local authority that rateable values had increased ten-fold in 1990.

By multiplying the rateable value of a property by the business rate applicable in each year a company occupied premises on an enterprise zone, a figure could be determined for the rates subsidy received by a particular company (see Appendices 4 and 5). On this basis, a figure of £6,870,115 could be determined as being a **conservative** estimate of rates subsidy to companies on this South Kirkby enterprise zone site, between August 1981 and October 1990. It is conservative because **some** allowance was included for rates subsidies to companies that had 'been and gone' — large

143

employers involved in double-glazing, textiles and chemicals that departed the zone, after a period of trading and collecting subsidies, for example — but by no means was allowance made for all the companies that had 'been and gone' in this way. The rapid rate of occupational turnover of industrial units rendered that a near impossible task. A detailed break-down of how rate subsidy levels were determined is contained in Appendix 5.

The figure of £6,870,115[1] represented only part of the total subsidy made available to firms on the enterprise zone , of course. It was estimated that rates subsidies accounted for 28 per cent of the net public cost of the enterprise zone experiment (PA Cambridge Economic Consultants, 1987). There was no reason to believe that the South Kirkby enterprise zone represented an exception to this. £6,870,115 is 28 per cent of £24,536,125. All that sum represented a subsidy direct from central government. Added to this should be a figure of £649,000 invested in infrastructural development at the South Kirkby enterprise zone by the City of Wakefield Metropolitan District Council, up to the end of 1987. It should be stressed that this investment by the local authority was made as a result of enterprise zone designation, and it was unlikely otherwise to have been made.

Given that the industrial estate had a pre-enterprise zone life stretching back into the 1950s, additional post-enterprise zone infrastructural investments can be legitimately included in cost per job calculations.

It should also be noted that the money used for infrastructure investment came from more than one source: £200,000 of it came from the European Community: £600,000 came in the form of an Urban Aid Grant from central government; the rest was local authority money (City of Wakefield Metropolitan District Council, 1990). The resultant combined subsidies figure of £25,185,125 remained a conservative estimate of

1. None of the figures used in this chapter have been adjusted for inflation: it would have unduly and unnecessarily complicated the financial calculations.

subsidies made available to companies involved in some way in the enterprise zone under consideration, between the period August 1981 and October 1990. Because of the difficulty of obtaining the information for the reasons outlined in Chapter 5, no account was made for the 'soft loans' (loans at below the normal interest rate) to companies by British Coal's job-creation arm, British Coal Enterprise. Nor was any account taken of money received by businesses from European Coal and Steel Community Assistance, thought it is known that some companies did take advantage of this, at least until its curtailment in 1982 (National Audit Office, 1986). The City of Wakefield Metropolitan District Council's Economic Development Unit also in some cases paid the first year's rent on small industrial units for small businesses on the South Kirkby enterprise zone; this too was ignored in the calculation of subsidy.

Additionally, English Estates, the government agency that worked on behalf of the Department of Trade and Industry had, as of October 1990, developed a total of 146,813 square feet of factory and workshop space, both prior to and after the designation of enterprise zone status on the industrial estate under consideration (English Estates, 1990). The public investment channelled in this direction — of providing advance factory units — was therefore of considerable magnitude, but was ignored on the grounds that English Estates claimed to have sold some of the units at a profit, and/or been in continuous receipt of market rents payable by occupiers.

What, then, in employment terms, did the £25,185,125 industrial subsidy to private companies achieve in buying?

A postal (documented here as Appendix 1), telephone and visual survey conducted by the author in the period August to October 1990 revealed a total of 57 companies operational in the South Kirkby enterprise zone at the time, providing an equivalent of 2,120 full time jobs.[1]

1. One company operated only in the three month run up to the Christmas period, employing about 100 people for that time, and six for the rest of the year. This was classed as being equivalent to 32 full-time jobs. Another company, the second largest employer

145

Figures collected by the City of Wakefield
Metropolitan District Council showed a total of
19 companies, providing 1,215 jobs, in September
1981, just after the designation of the
industrial state as an enterprise zone.

Subtracting the 1,215 from 2,120 jobs, this
meant that £25,185,125 'bought', at best, 905
jobs. From this figure, however, had to be
deducted the number of jobs that could be
identified as being merely local relocations,
where companies moved to the zone from very
nearby simply to take advantage of the absence
of business rates, or because the zone presented
the only industrial land that was available and
companies were looking for premises in the
locality anyway. Companies in the former
category — transferring from nearby in order to
take advantage of the absence of business rates
— are designated 'l.r.' in Appendix 5: local
relocation. In the case of local relocations,
or firms that would have been in the locality
anyway, it can be asserted with some confidence
that the paid employment provided by these two
categories of companies would have been similar
in both quantity and quality even if the
enterprise zone had not existed.

The biggest employer on the estate at the
time, moving on the zone in October 1990, was a
mens' outfitter, providing 370 jobs (Company 1
in Appendix 5). But upon moving to the zone,
the company closed one factory in South Kirkby
itself and one three miles away, where the
company were occupying the former baths of a
redundant pit. Such premises were clearly
unsuitable for a garment manufacturer, and the
company was understandably seeking an
improvement. Nevertheless, this local
relocation involved neither an increase nor
decrease in the number of jobs available. Thus
the figure of 370 had to be deducted from 905,
leaving 535 'new' jobs.

A chemical company was another relocation,
starting life on the zone with 40 jobs but, by
October 1990, employing 70 (Company 9 in
Appendix 5). Logically, then, at least 40 had

in the zone at the time, said employment varied between
200 and 300. For the purpose of this chapter, employment
was taken to be 250.

146

to be subtracted from the figure of 535, leaving 495 'new' jobs attributable to the zone itself. The list of local relocations went on, however, and included: a glass recycling company (Company 16 in Appendix 5); a security company (Company 21 in Appendix 5); a Christmas hamper packing company (Company 18); a joinery company (Company 25); an engineering company (Company 20); a tea distribution company (Company 28); amongst others.

Of the 57 companies trading on the enterprise zone as of October 1990, 15 could be positively identified from the survey as being local relocations (companies 1,9, 14, 16, 18, 20, 21, 25, 28, 39, 41, 44, 45, 50 and 57 in Appendix 5). Associated with these companies were 579 'transferred' jobs (this was not the same figure as the total number employed by these companies; for explanation, see the comments in the last paragraph on the chemical company). So from the figure of 905 jobs arrived at earlier, 579 had to be subtracted as representing merely locally transferred employment.

This left a total of 326 'new' jobs that had some claim to being there as a consequence of enterprise zone status being granted to Langthwaite Grange Industrial Estate.

£25,185,125 divided by 326 meant that the cost per job to the public purse — largely the British taxpayer — was £77,255.

Again, it should be stressed that this was a highly conservative estimate. Many of the companies that were on the enterprise zone would have been there anyway, but the jobs they provided were counted here as 'new' jobs. Many of the smaller businesses, for example, were run by local people who lived in the area anyway. A mining engineering firm (Company 8 in Appendix 5), one of the largest employers on the zone, said it had moved its headquarters from South Wales to the South Kirkby enterprise zone to be near what was left of the coal mining industry. The bulk of its operation had in any case been on the industrial estate prior to its designation as an enterprise zone.

Of the six employers on the zone employing over one hundred people in October 1990, three of them had been on the industrial estate anyway, prior to its change in status. Two of these were frozen food distributors (Companies 2 and 3 in Appendix 5), the third a sports shoe manufacturer (Company 5 in Appendix 5). Another of the six was a local relocation, the gents' outfitter (Company 1 in Appendix 5) referred to above.

One of the frozen food distributors — Company 2 in Appendix 5 — was a subsidiary of the Swedish multinational AGA AB through Frigoscandia AB (Sweden). To the year ended 31 December 1990 the total British operations (they had several) made a gross profit of £7,231,000 (Extel Financial Vanguard Companies Service, 1991). A case can be argued that this was hardly an equitable use of an estimated £2,406,074 taxpayers' subsidy over the period in question. At least 34 companies operating on the zone as of October 1990 were either on the industrial estate prior to enterprise zone designation, or were positively identified from the survey as companies which would have been on the industrial estate or very near, enterprise zone or not: companies 1, 2, 3, 4, 5, 8, 9, 10, 12, 14, 16, 18, 19, 20, 21, 23, 25, 27, 28, 30, 32, 37, 39, 40, 41, 44, 45, 46, 49, 50, 51, 52, 53 and 57 in Appendix 5. Windfall gains, local relocations, or companies that would have been in the immediate locality anyway then, accounted for a minimum of nearly 60 per cent of the companies trading on the estate as of October 1990. The same companies accounted for 1,641 jobs out of the 2120 total on the estate; over 77 per cent. Hardly an encouraging sign of new enterprise and new endeavours.

By way of comparison, the South Kirby-Riddings pit complex, which closed in March 1988, was said to have lost — and therefore cost the taxpayer — £63.5 million since April 1982. For the purpose of this study, this figure had to be accepted as an accurate reflection of the operating costs of that colliery, though recognition should be made of the fact that the accounting techniques employed by the National Coal Board/British Coal have been subjected to severe criticism (Berry et al, 1985; Robinson, 1984).

In broader terms, it has been argued that the cost to the taxpayer of closing an 'uneconomic' pit, because of costs of redundancy, social security payments necessary, subsidence compensation payments that continue to be made after coal production has ceased, lost rent and rate revenue to local authorities, was higher than the cost of keeping the pit open (Glyn, undated; O'Donnell, 1988). The reply of the free-marketers was to argue that closing loss-making pits would actually result in a net creation of jobs in the economy; closing 'uneconomic' would mean — at least theoretically — that taxation could be cut, which would lead through to a higher level of employment in the economy as consumers' spending power increased. Closing uneconomic pits would also free labour to work in productive expanding sectors (see, for example, Minford, 1984).

So, according to this argument, via changes in taxation and demand structure, and the provision of cheaper coal, closing 'uneconomic' pits would actually lead to a higher level of employment in the economy. This was so even in high unemployment, mining areas. Minford and Kung (1988) argued that:

'There is no evidence that on balance the new jobs in high unemployment areas will not match up with the job losses.'

At the outset, it would have appeared as if the figures for subsidies of £63.5 million for South Kirkby-Riddings pit and £25,185,125 for the enterprise zone were in different leagues. But in terms of **proportionate** economic activity generated or maintained, and resultant employment creation/maintenance, they were not.

South Kirkby pit was merged with Ferrymoor-Riddings colliery in March 1987, to become South Kirkby-Riddings. Employment levels at the pit complex varied. In 1983, the then two pits employed 2,111 people (Barnsley Metropolitan Borough Council, undated). At the end of the financial year, employment at South Kirkby pit was down to 1,300 leaving an estimate combined total at South Kirkby and Ferrymoor-Riddings of 1,849 (*Hemsworth and South Elmsall Express*, 5

June 1986). By March 1987, employment at the now merged pit complex had fallen again, to around 1,599 (*Hemsworth and South Elmsall Express*, 16 March 1988). By May 1987, the pit complex employed something over 1,350 (*Hemsworth and South Elmsall Express*, 27 May 1987). Taking the four employment figures between 1983 and 1987 for the pit complex, summing them together and dividing by four, it would seem reasonable to argue that the pit complex provided an average total of 1,727 jobs between 1982/3 and closure in March 1988. If that figure was accepted, it meant that the cost per job to the taxpayer for the coal mine jobs was £36,769 (£63.5m divided by 1,727). This compared to £77,255 for the 'new' jobs on the enterprise zone. Moreover, £36,769 was an over-estimation of the cost per job because it ignored payments made back to public authorities by British Coal in the way of business rates.

The time periods over which this comparative study ranged were, of course, slightly different. In the case of the pit, ranging from April 1982 to March 1988; in the case of the enterprise zone, from August 1981 to October 1990. Any attempt at further differentiation, however, into say cost per job per year, would have been an almost impossible task: it was not possible to determine precisely how many jobs had lasted for exactly how long on the enterprise zone. The 370 jobs at the gents' outfitters were a case in point: at the time of the survey, October 1990, they were completely new to the zone, though not to the locality.

A central point of this study, then, was this: the cost per job to the public purse was higher in the Conservative government-sanctioned and initiated enterprise zone than the cost per job involved in keeping open in 'uneconomic' pit.

The case presented here, however, is not that all 'uneconomic' coal mines should always be subsidised, whatever the cost. That would be an untenable position which would ignore the fact of scarcity of resources in the economy. In the absence of an economy of total abundance, there has to be established a hierarchy of needs in society, and demands made on the government would be met in line with that hierarchy where resources allowed. A social and political mechanism becomes necessary in order to determine this hierarchy and, in practice, is in operation all the time as international, national and local governments decide which projects, groups, individuals, firms, to support in the face of competition from others (see chapter 2).

The argument here can be summed up, rather, by arguing that: public money, from a variety of sources, was used to subsidise private firms in this particular case. The 'opportunity cost' of doing that was that the money could have gone to an alternative job creation/maintenance scheme, which might, or might not, have involved subsidising the local coal mine.

Cost per job and cost to the public purse, of course, were not the end of the matter. It was noted earlier that there was a need in the locality to change the industrial structure of the area. There was a need for the economic base to be diversified. It had, for far too long, been dependent upon one industry, coal, which in the 1980s and early 1990s was in rapid decline. Was that industrial structure changed by the implementation of the enterprise zone policy?

In fact, outside the coal mining industry, the industrial structure of the locality in 1990 was much the same as it was in 1981, when the enterprise zone was started. The biggest employers on the industrial estate before its designation as an enterprise zone in 1981 were frozen food distributors. Textiles and sports footwear were also relatively big employers in the locality. By 1990, they had not changed.

Hall foresaw enterprise zones as becoming 'independent centres of technological innovation'. As of October 1990, the enterprise zone had been in operation over 9 years and there was no sign of what Hall forecast: there was nothing whatsoever approximating to high-technology industry.

Hall's other claim was that enterprise zones would encourage 'enterprise': if enterprise could be judged as being related to levels of economic activity, then what additional enterprise there was limited and expensive to the taxpayer.

What **was** in the process of change in the locality, it could be argued, was the industrial **culture**. Consider the situation. The coal industry in Britain had been in public ownership since 1947; it had traditionally been 100 per cent unionised; it had a reputation for worker militancy; it was a large scale industry, organised in large scale production units.

In terms of labour organisation, Massey (1983) noted 'the National Union of Mineworkers was a leading union in the fight for the closed shop, and was one of the first to win recognition of this principle from employers'.

As far as worker militancy was concerned, it has been estimated that from the 1930s to the 1950s, miners, accounting for just 4 per cent of the economically active population, were responsible for over one third of lost time due to industrial disputes in Great Britain (Beacham, 1958). The 1970s and 1980s saw three major national strikes by the National Union of Mineworkers — in 1972 and 1974, over pay; in 1984-5, against pit closures — on each occasion fought while the Conservatives were in governmental office.

Clearly, the Thatcher government would not be ideologically enamoured of the idea of supporting this industry with financial assistance. There was a sense in which it could be argued that the coal industry in Britain represented in microcosm everything that they saw as being wrong with Britain's industrial economy. It was in public ownership; it was heavily unionised with a tradition of at least unofficial worker militancy, and union-sanctioned worker militancy in the 1970s and 1980s; it had frequently not managed to support itself financially, and had therefore had to call for support from government.

This, of course, stood in stark contrast to the economic activities on the enterprise zone under consideration. Firstly, **all** of the businesses on the enterprise zone were privately owned. Secondly, many of the businesses were very small scale: 35 of the 57 companies, or nearly 62 per cent, located on the zone as of October 1990 had 10 or less employees; another 9 companies employed 40 or less, meaning 44 companies, or 77 per cent of all the companies on the enterprise zone, employed 40 or less people.

A third major difference between the coal industry and the companies on the enterprise zone related to unionisation. In 43 of the 57, or 76 per cent of the companies on the enterprise zone, there was no trade union organisation. It could be argued that this was simply a reflection of the fact that many of the businesses on the enterprise zone were small scale, but non-unionisation was evident even in some of the bigger operations on the zone. A meat-packing plant employing 113, for example, had no union; a condom distributor employing 35 had no trade union.

Elsewhere, evidence suggested 'weak' trade unionism. At a mining engineering company employing 87 people, for example, employees belonged to a mining management union — British Association of Colliery Managers (BACM) — not affiliated to the TUC.

The economic sector that emerged in South Kirkby enterprise zone neatly fitted three prongs of the Thatcher government's industrial policy strategy: anti-trade unionist; pro-private ownership; a reliance on small and medium-sized firms to mop up unemployment and to give a boost to areas of the economy hit by industrial recession.

Despite the fact that the outcome was a cost to the public purse per job higher than would have been necessary from subsidising a coal mine, this study clearly showed that industrial subsidies were available from the Conservative central government of this period in the ideologically 'correct' context. And the ideologically 'correct' context, as far as the Thatcher government was concerned, was where an industrial culture could be manipulated away

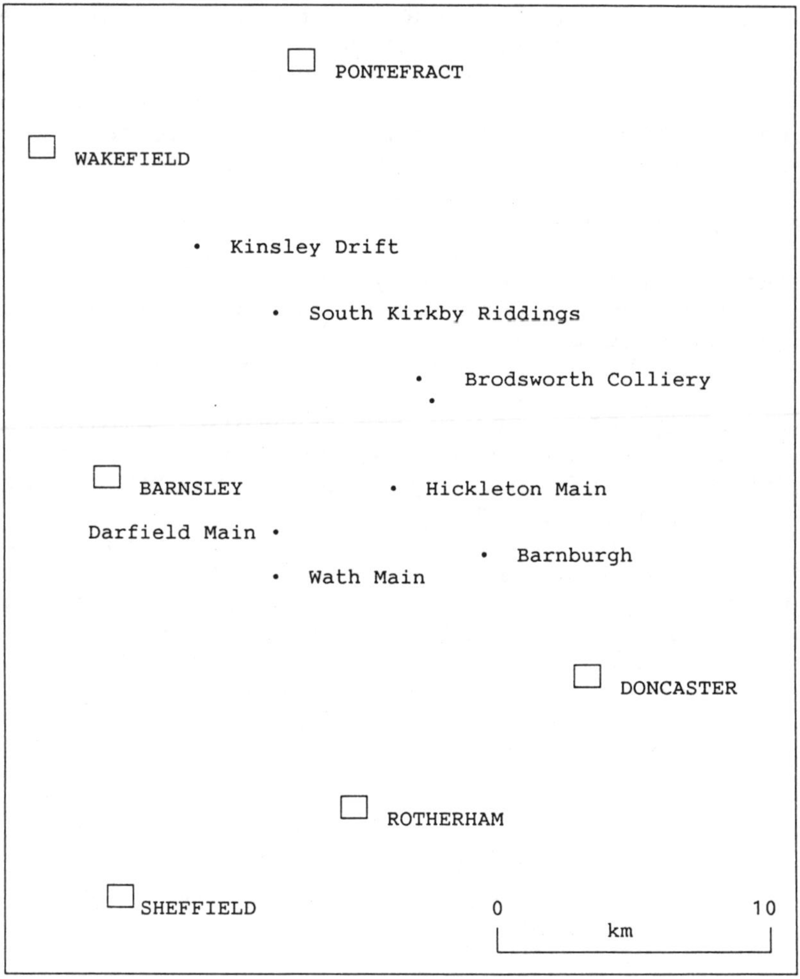

Figure 1 Location of collieries within 5 miles of South Kirkby

from public sector ownership, high level of trade union organisation, high level of worker militancy, to the opposite of these. Bringing about such change would have been seen as being a positive use of funds.

One expression indicative of these 'industrial culture' arguments, and indeed suggesting that the 'old' coal-mining industrial culture had survived, came from a company which had relocated to the zone from a nearby city, a non-mining location. No claim is made here as to how representative this was of other employers on the zone, but the comments of this company were probably worth recording. In complaining about the workforce, the company said:

> 'They are not reliable or company orientated. Perhaps due to previous broken promises from the Coal Board regarding job security. They have a very short time frame in almost everything we do for them.'

Even where it could be shown that subsidies to companies on an enterprise zone outweighed subsidies to nearby coal mines in cost-per-job terms, there would still be those who argue that subsidising enterprise zone companies was more economically justified than subsidising a coal mine.

Subsidising the coal industry, they would argue, was propping up a dead industry, a thing of the past, like trying to breathe life into an industrial ghost.

Subsidising firms in an enterprise zone was helping to start off something new and dynamic, where subsidies would cease after a period and a thrusting, dynamically-orientated sector of the economy would rise like a phoenix from the ruins of the old.

All that can be said was that, in practice, in terms of the outcome of policy, this did not happen. There was little economic activity generated. Unemployment in the South Kirkby region was at 9.2 per cent in July 1990, compared to a national figure of 6.6 per cent in July 1990 for the UK as a whole. The figure of

9.2 per cent was considered to be a significant **underestimation** of the true level of unemployment in South Kirkby, however, as it measured only those who were claiming unemployment benefits.

There were many more, as a consequence of the Redundant Mineworkers Scheme, which, since 1989, did not require redundant mineworkers over a certain age to sign on as unemployed to receive benefit (North Derbyshire Coalfield Partnership, 1990), who were, nevertheless, effectively unemployed or, alternatively, early retired. The true figure of unemployment in South Kirkby as of October 1990 was probably closer to 18 per cent than 9.2 per cent.

So the industrial structure of South Kirkby was not changed during this period by the enterprise zone. There was nothing on the enterprise zone that could be classed as high-technology industry, laying the basis for future economic expansion. The economic base was not successfully diversified. British Coal remained the largest employer in the area as of October 1990.

There was still a school of thought, nevertheless, which argued that subsidies to industry could be divided into those which would help to form a 'positive' adjustment to an economy, and those which formed a 'negative' adjustment. Under 'positive' adjustments, market signals would be followed through to their logical conclusions. So, for example, a coal mine would be closed if it did not have a market for its coal. A 'positive' adjustment in this case might be a government paying subsidies to prolong the life of a coal mine, so that the social costs imposed on a community by a pit closure could be spread over a longer time period than would otherwise have been the case. Another example of a 'positive' adjustment might be where a government would finance a retraining programme for redundant workers.

As the OECD (1979) Council, meeting at Ministerial level, agreed in 1978, however, the ultimate objective of positive adjustment policies was to 'phase out obsolete capacity and re-establish financially viable entities'.

In a broader definition, the OECD (1984)

characterised positive adjustment measures as marked by efforts:

'(i)   to encourage the shift of labour and capital to activities in line with their comparative advantage and relative prices in keeping with international competitive developments, new technologies and altered patterns of consumption and production;

(ii)   to remove barriers and measures which inhibit the movement of existing employment and productive capacities to more productive uses; and

(iii)   ease rigidities in financial and labour markets which raise costs and impede the full and efficient use of resources.'

A 'negative' adjustment, on the other hand, was where subsidies were made by a government to defend a status quo position, or to preserve inefficient industrial and economic structures. Advocates of pit closures in the 1980s and 1990s argued that to subsidise loss-making coal mines would have represented a 'negative' adjustment. Such subsidies should therefore not be made.

The 'positive' and 'negative' argument, however, could only be valid if two conditions were agreed to.

Firstly, the argument assumed that the market was always the most efficient way of deciding what should and should not be produced in an economy, and of deciding who should be employed doing what. Those seeking to prevent pit closures in the 1980s argued that the market was not always the best means of doing that. This was so, firstly, because coal was a finite commodity, and therefore could not be treated in the same way as other, 'renewable', commodities: £5 worth of oil was not the same as £5 worth of wine. Once the wine was consumed, it could be reproduced from grapes; the oil could not be reproduced, therefore greater care in the husbandry of finite resources was called for (Kirk, 1982).

In the early 1990s, the argument against allowing the 'market' to dictate which pits would or would not stay open took on a new dimension. For, particulary from October 1992 onwards, it was being argued by many political and economic commentators that the market itself was rigged in favour of electricity generation from gas and nuclear power, and against the generation of electricity from coal (Coyne, 1992). This meant, according to the argument, that not even a market economist could trust the market as it was constructed as of October 1992 to influence decision making in relation to energy policy.

The second condition that needed to be agreed to if the 'positive'/negative' argument was to have any validity, was that it did not matter if the entire British deep-mining industry closed down. Again, those seeking to argue against pit closures in both the 1980s and the early 1990s put forward the case that to close down large parts of the coal industry in Britain would leave the economy unhealthily reliant on energy producers abroad. In some cases, the energy producers were in politically and economically unstable countries, which might further endanger supplies. A parallel was drawn between the British economy's potential position in the 1990s and beyond — where it looked likely, at least until October 1992, to become reliant on coal imports — and its position in the 1960s, when it was reliant on oil imports from the Middle East. The latter ended disastrously after the large-scale oil price hikes of the 1970s.

Further, it was not always clear in the 'positive/negative' debate exactly what was 'positive' and what was 'negative'. Was it more 'positive' to subsidise a frozen food distributor owned by a foreign multinational that was already operating in an industrial estate prior to its designation as an enterprise zone, than a coal mine producing finite energy resources which companies and individuals in the economy might need at some stage in the future, even if they did not need them in 1990?

Even the OECD, with its emphasis on market-orientated policies, recognised that in the energy field, there might be a need to follow practices which overrode market signals:

'There are certain areas where markets are unlikely adequately to reflect and anticipate future economic and social needs. This applies, for example, to research and development in producing and saving energy; to improvements in environmental quality, health care, urban infrastructure etc' (OECD, 1979).

6.5 Conclusion

In contradiction to its frequently expressed preferred free market stance on economic and industrial policy, the post-1979 Conservative government **was** prepared to provide subsidies to business and industry. The condition of such provision was that those subsidies served **ideological** objectives.

This case study demonstrates that subsidies to private, very often small scale, businesses on an enterprise zone in a coal mining locality were considered politically 'sound', whilst continued subsidies to the coal industry were not.

The political environment fostered in the Thatcher years emphasised that the ability to make profits should be the central criterion of whether or not a company, business or industry survived. All else was subjugated to this deification of 'economic realism'. Hence the economic devastation visited upon pit towns throughout Britain by deindustrialisation in coal mining.

'Economic realism', as defined here, did not have an exclusive legitimacy as a perspective for shaping and prescribing economic, social or industrial policy, however. A case could be made for saying that an equally legitimate approach to these policy areas was to put the needs of people and communities as a major priority. From that perspective, in terms of the economic welfare of the residents of South Kirkby, there was no justification for the political decision to subsidise enterprise zone companies rather than the pit. Insufficient numbers of jobs for the community were created by the enterprise zone during the period under review to replace those lost in coal mining, for instance.

Even from a more tangible, public expenditure, viewpoint, there was little, if any, sense in the decision to subsidise enterprise zone companies as opposed to the pit. The cost-per-job of the subsidising enterprise zone companies was far higher than the cost-per-job of subsidising the coal mine.

In the absence of economic logic, the conclusion has to be that policy motives were exclusively governed by an adherence to a purely ideological viewpoint. Subsidies to enterprise zone companies, rather than to the coal industry, bolstered a central objective of post-1979 Conservative government industrial policy. In short, the 'industrial culture' of the local economy was manipulated **away** from the despised nationalised, large scale, unionised industrial sector and into a preferred non-unionised or weakly-unionised, privately-owned, small and medium-sized business and industrial sector.

For the people of South Kirkby and similar pit towns, however, the price of this determination to stamp out the old and attempt to replace it by the new was undoubtedly very high.

The methodological tools of analysis employed here were an interview in 1990 with the deputy leader of Doncaster Task Force; liaison with Doncaster Metropolitan Borough Council's economic development unit; a telephone interview with a leading official of Doncaster Chamber of Commerce and Industry; liaison with the Yorkshire Area of the National Union of Mineworkers on labour force levels at pits; and written communication with the Craven Tasker trailer-manufacturing company. None of the individuals or organisations mentioned were responsible for the interpretation of information presented here; some might disagree with it. Subsidiary information was taken from the local press, minutes of the meetings of Doncaster Metropolitan Borough Council, and some of the literature on inner-urban policy.

7.2   The Task Force idea

Task Forces with some resemblance to the kind later employed in Doncaster and elsewhere were first initiated in Scotland by the Scottish Development Agency, at the instigation of the Scottish Office. Two were established, each with a life span of five years, in Garnock Valley and Clydebank, in 1979 and 1980 respectively. They were responses to rapid declines in the local economies concerned, especially the closure of the Singer works in Clydebank, and the closure of the British Steel Corporation's steel-making plant in Garnock Valley. Both localities had been heavily dependent, economically, on the two closed works (Gulliver, 1984).

In England, the origins of the Task Force concept lie in the 1981 riots in Toxteth, Liverpool. The first Task Force, initiated as a support to Michael Heseltine, temporarily the unofficial 'Minister for Merseyside', was a policy response to those riots and the problems of urban deprivation that contributed to their cause (Lawless, 1989).

Heseltine's Task Force in Liverpool consisted of about 30 full-time civil servants. Most of these were drawn from Manchester offices

163

of the Department of Environment, though there were some secondees from the Department of Industry and the Manpower Services Commission. It also had managerial secondees from the private sector, up to 15 in its first year, falling to five in its second (Parkinson and Duffy, 1984).

Essentially, the Task Force approach was based on a strategy of co-ordination: the bringing together of representatives of business, local authorities and central government to ensure the implementation of specifically targeted attempts at economic regeneration.

As the Conservative government document, **Action for Cities**, had it:

'They work directly with local business, local people and local councils, and act as broker between the public and private sectors ...

'Pooling the resources of the private and public sectors is the way to achieve real success in the inner cities' (HMSO, 1988).

Heseltine's Task Force was 'project orientated' in its first phase, concentrating on highly visible schemes, such as the reclaiming of derelict land in conjunction with Plessey to form the Wavertree Technology Park. This was aimed at attracting high technology business to Liverpool.

The Thatcher government quite clearly became keen on the Task Force idea. Eight were initiated in the summer of 1986; eight more in 1987. The Liverpool Task Force was the model. In 1988 the Thatcher government was publicly proclaiming their success:

'So far over 350 local projects committing almost £13 billion have been supported; about 7,000 businesses have been supported or started, with 1,700 new jobs; over 200 companies are involved, contributing around £4.5 million' (HMSO, 1988).

The second phase of Task Forces were on a much smaller scale than Liverpool. Doncaster, for example, had six full-time civil servants all seconded from the local branch of the Department of Employment plus, for a year, a British official from the CEC. Importantly, there were no managerial secondees from the private sector. This might have reflected the ambiguity as to the role managerial secondees were meant to play in Task Forces (Parkinson and Duffy, 1984), as well as the fact that the post-Liverpool Task Forces were not such high profile operations as the one in Merseyside born out of inner-city riots.

Nor, unlike Liverpool, was the Doncaster Task Force inter-departmental in terms of central government representation. This original inter-departmental approach had been intended to be a response to a lack of co-ordination on inner-urban policy, where the Department of Environment would be pursuing certain projects, whereas the Department of Employment would be pursuing yet others without any interaction. Co-ordination, it was believed, would improve the impact of inner-urban policy. But, in any case, it appeared as if inter-departmentalism had mixed results in the Liverpool case. Whilst on an individual level, Task Force members benefited from the mix of expertise, on the other hand, different government departments continued to have different approaches to inner urban problems (Parkinson and Duffy, 1984).

But the single-department nature of the Doncaster Task Force probably reflected practicality and pragmatism rather than anything else: every town and city had a Department of Employment that staff could be drawn from. That was not the case with the Department of Environment or the Department of Trade and Industry.

7.3 Objectives

An assessment of the effectiveness of a policy is best attempted by relating eventual policy outcome to a given set of initial policy

objectives   This is the evaluation criteria
which has been applied here. An important point
about public policy objectives, however, is that
they are often vague, ambiguous, unclear (Burch
and Wood, 1983). They are often 'pliable', so
that in any after-the-event policy evaluation,
the objectives can be bent, can be made to
'accommodate' the eventual policy outcome.

The objectives of the Task Force initiative
generally, often referred to as the four 'E's',
proved no exception to this general rule. The
objectives were: to provide employment for
local people;   to encourage and facilitate
enterprise by local people;   to enhance the
'employability' of local people;   to support
projects designed to improve the physical
environment.

Within these general guidelines, each Task
Force produced 'action pans' which outlined how
specific objectives would be achieved.

Doncaster Task Force produced four such
action plans: in October 1987, in January 1988,
in 1989, in 1990. As of January 1991, when this
research was carried out, the specific content
of these remained confidential.   A broad
outline, however, was available.

The first action plan was comprised of the
results of research into the economy of the
locality.   The coal industry had been
traditionally a major employer in the area, but
outside this nationalised sector, the Task Force
found that the bigger employers were largely
'satellite' companies of bigger multinational
concerns, such as ICI, manufacturing fibres and
paints;   and Case, a tractor-manufacturing US-
multinational.   The Task Force believed that
proportionately, Doncaster was under-represented
in relation to the national economy in terms of
medium-sized employers like these.   The Task
Force believed that, in the economic climate of
the late 1980s, increased employment would not
be likely to be created by these 'satellite'
companies which, if they were to expand at all,
would expand in their home bases.

According to the Task Force, however, their
research did discover that Doncaster had a
significant small business sector. Apparently,

166

in the late 1980s, Doncaster had more than 6,000 companies that could be categorised as being part of the small business sector, 95 per cent of these employing less than 50 people. For a locality such as Doncaster, with its 'traditional' industrial structure based around coal mining and manufacturing, this was considered by the Task Force to be proportionately higher than had been expected.

Indeed, in terms of the policy thrust of the creation of an 'enterprise culture' in the 1980s, where a large part of the definition of an 'enterprise culture' was concerned with the propensity of people to set up businesses, and the existence of a substantial small business sector (Turner, 1990), Doncaster's high rating here was perhaps surprising. Working class areas reliant on traditional industries such as coal and engineering did not usually exhibit in the 1980s a buoyancy in terms of 'enterprise culture'.

This proportionately large small business sector persuaded the Task Force to address itself to that part of the local economy, on the basis that expansion was at least a possibility in that sector, and therefore extra employment might arise there. Part of the Doncaster Task Force's efforts, then, were devoted to organising business advice to the small business sector: advising companies within the sector, for example, where and how they could get access to loans and grants. In the words of the deputy Task Force leader, these small businesses needed 'job creation capacity building into them'.

An important finding of Doncaster Task Force's first action plan was that 'the Task Force area, although suffering in most indices, does not stand out in stark relief when compared with the surrounding area'.

Indeed, while the Task Force itself reported that it had had a good working relationship with the local authority — which contrasted with Liverpool, where there appeared to have been tension (Lawless, 1989) — one point of conflict was over the boundaries of the Task Force area.

The leader of Doncaster Council in 1987,

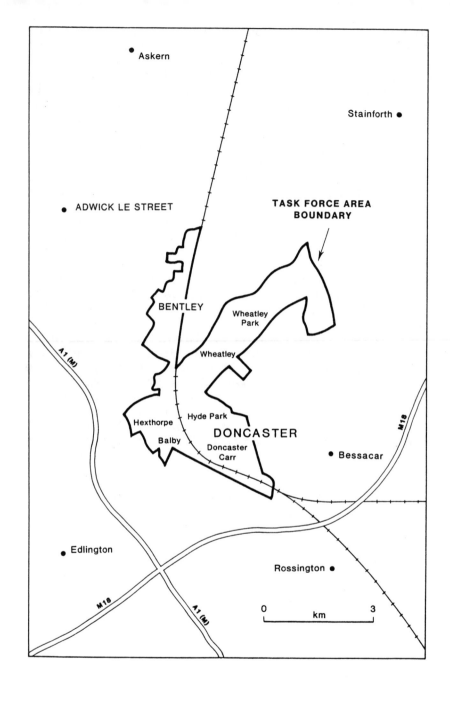

Figure 2: Doncaster Task Force boundaries

Gordon Gallimore, expressed disappointment at the lack of consultation between Whitehall and the local authority on the territorial boundaries of the Task Force.

Gallimore responded to central government's announcement in April 1987 that a Task Force would be established in Doncaster, by saying:

> 'We put forward the hard hit villages of Askern, Stainforth and Rossington, but they were ruled out. We also suggested Edlington, where a pit has closed down. The out of the blue came the present proposal.

> 'No opportunity has been given for us to influence the choice, and that has left a slightly bitter taste' (*Doncaster Star*, 28 April 1987).

What the Task Force in Doncaster, and in the other areas, did represent, was an attempt to by-pass local authorities in the localities, and in the policy areas, that the municipal authorities might at one time have considered their exclusive preserve. Urban policy under the Conservatives from 1979 to the time of writing, the early 1990s, was characterised by this attempt to by-pass local authorities, whether it be through enterprise zones, urban development corporations, Task Forces, City Action Teams, or other policy mechanisms. (Stewart, 1987). The position of the Conservative central administration in power since 1979 appeared to have been that local authorities were simply not to be trusted or were inefficient, in relation to fostering the changes 'necessary' in local urban economies (Haughton and Roberts, 1990). In the localities most in need of urban regeneration, the majority of local authorities were Labour controlled, deepening the above two 'problems' perceived by central government. The disregard of Doncaster Metropolitan Borough Council's views on where the Task Force area should be situated represented further evidence of the by-passing of local authorities.

169

In any policy that involved targeting, as the Task Force policy did, there were bound to be definitional problems. Who could be 'defined' as deserving beneficiaries of policy? And how?

As pointed out by MacLachlan in his discussion of Townsend's concept of poverty, definitions of poverty are neither true nor false, and are essentially value judgements, unlike factual claims (MacLachlan, 1983).

And as Brown and Madge (1982) noted in their report on the ten year Department of Health and Social Security and Social Science Research Council programme of research into transmitted deprivation:

'What emerges from this [report] is a highly fragmented notion of deprivation. The term is used in so many ways that it has become almost meaningless. Thus the only feasible approach for this report ... is to accept ... that there is no single state of deprivation or disadvantage.'

The Task Force policy was essentially an exercise in area-specific positive discrimination. As such, it represented a continuity in urban policy initiatives reaching at least as far back as the Urban Programme, originally instigated in 1967, and, more tenuously, to the development of 'specific area' policies as a consequence of the Barlow (1940), Scott (1942), and Uthwatt (1942) committees (Edwards and Batley, 1978).

In some Task Force localities, this exercise was easier than in others. Chapeltown, in Leeds, for example, clearly represented an area of deprivation in a wider city context of relative prosperity. The same could be said for St Paul's in Bristol, site of another Task Force. In Doncaster, this was not as clear. Much of the borough exhibited similar social and economic characteristics as the Task Force area.

The decision regarding within which territorial area residents would receive Task Force benefits was decided by civil servants at the Department of Trade and Industry, using a multi-deprivation index, as it was in all Task Force areas. The second phase Task Force programme was a DTI-led initiative.

It should be acknowledged, however, that the determination of the multi-deprivation index reflected the **values** and **value** **judgements** of those drawing it up. Hence, the decision on who was to benefit from policy targeting such as this was also dependent upon value judgements (Brown and Madge, 1982). Who benefits and who does not benefit is not a product of neutral arbitration.

The multi-deprivation index was drawn up with specific references to **inner** **city** deprivation. Doncaster, however, was not what was usually taken to be an inner city locality. It was not a Toxteth within a city. It was a free-standing town that during the time of writing was suffering from deindustrialization in traditional sectors of its local economy: most notably coal mining, but also railway engineering, and other sectors.

The Task Force's second action plan followed in January 1988. Basically, this was a revision of the first, but incorporated more detailed objectives and priorities. Key among these was the identification of Doncaster Carr, an industrial estate within the Task Force area which had a surplus of undeveloped land, as a site for developing a 'training park', managed workspace and business starter units. Another objective of the plan was to rejuvenate the Chamber of Commerce and to organise the promotion of local purchasing by the bigger firms in the area.

By the time of the third and fourth action plans, in 1989 and 1990, the emphasis that had emerged related to 'enterprise' and 'employability'. The objectives of job creation and environmental improvement had apparently proved more difficult to stimulate.

The fourth action plan explicitly recognised that Doncaster was different from other Task Force areas. The Task Force, as was argued earlier, had not been given a deprived locality within an area of greater relative prosperity, to work within. And so, unlike in some other Task Force areas, the plan would not be to divert resources and benefits from more prosperous parts of the locality. Instead, the Task Force would attempt to ensure that there was economic regeneration in Doncaster, and that at least some of the benefits of this would reach Task Force area residents.

7.4 The locality

An examination of the locality within which the Task Force was operating serves to outline the nature and extent of the problems that faced it. Dealing with those problems would not have been an easy task for any public sector organisation, or indeed any other organisation, especially one on a restricted budget. That was not to argue, of course, that the Task Force had sole responsibility for solving those problems.

One of the lessons of the decision by central government to locate a Task Force in Doncaster related to an age-old part of the political craft: the potential effectiveness of lobbying. Other localities in the region, indeed other declining coal mining localities, for example Barnsley, were not allocated a Task Force, despite similar problems of deindustrialisation. The 18.5 per cent unemployment in the Doncaster Travel to Work Area in July 1987 was not much higher than the 17.2 per cent unemployment in the Barnsley Travel to Work Area at the same time, for example (Doncaster Metropolitan Borough Council, 1987).

Early in 1987, when Doncaster was facing deindustrialisation and consequently rising unemployment, from coal mine closures, the closure of British Rail engineering works, and retrenchment among other major employers, Doncaster Metropolitan Borough Council lobbied Lord Young. At the time Lord Young was Secretary of State for Employment. One objective of this lobbying of central government was to obtain a change in the designation of the Doncaster locality from Intermediate Area status under regional policy, to the designation of Development Area status. Achieving the latter would have improved the eligibility of companies in the locality for governmental financial assistance under the form of regional policy operating at the time.

Not surprisingly, given the diminution in the importance of regional policy under the Thatcher governments (Turner, 1989), and the policy trajectory that central had set for itself (Department of Trade and Industry, 1988), this request was not granted.

Nevertheless, Lord Young was apparently impressed by the case put forward on Doncaster's behalf, and by the commitment and enthusiasm shown by the councillors and officials who had lobbied him. And so, in keeping with the managerialist policy ethos of the times, Doncaster was offered, and accepted, a Task Force.

The Task Force's area in Doncaster incorporated varying proportions of five census wards. In a less bureaucratic definition of its territory, it was comprised of six recognised communities: Bentley, Balby, Hexthorpe, Wheatley and Townfield, Hyde Park, and Central Doncaster.

Bentley is a locality that has been traditionally reliant on the coal mining industry. Many at Bentley worked at Bentley colliery, still in operation as of November 1991, or had worked at the nearby Brodsworth colliery, closed in September 1990, or indeed at other pits nearby.

The past prosperity of Balby also rested on traditional industry: in particular engineering, related to British Rail, or the coal mining industry. Both suffered a massive rundown in the 1980s.

Hexthorpe's wealth traditionally relied almost exclusively on British Rail engineering workshops. The rationalisation of that sector resulted in substantial redundancies in the 1980s, through the railway engineering sector remained in 1990 a significant employer in the locality.

Wheatley and Townfield suffered from declining physical stock in its housing, high and long-term unemployment and general economic deprivation. Hyde Park suffered similar problems, compounded by apparently extensive drug and alcohol abuse among some residents. Central Doncaster was obviously the commercial/retail centre. Apart from that it had a limited amount of high-rise and council property, which shared some of the environmental and social disadvantages of many such inner-urban developments.

The Task Force covered a population that was calculated as standing at 43,000 in 1981. The Task Force estimated, however, that the actual population in its area stood at around 35,000, equating to 14,000 households. This compared with a total population under the administrative jurisdiction of Doncaster Metropolitan Borough Council in 1990 of 291,600 (Municipal Year Book, 1990). So, on that basis, the Task Force covered about 12 per cent of the total population of the Borough.

7.5   The resources

Doncaster Task Force had a budget of £3.4 million, covering expenditure between July 1987 to June/September 1991, on 119 projects.

This was a small budget. If the total sum is divided by the 35,000 figure for residents of the Task Force area, it amounted to £97 per person, or £24 a year. If the Task Force territory population was closer to the 43,000 of 1981, these figures fall to £79 and £29 respectively.

Moreover, a case can be made that any resources brought by Doncaster Task Force were dwarfed by cut-backs in public sector budgets elsewhere. The Labour Party was claiming in 1987, for example, that Doncaster Metropolitan Council had lost £73 million in rate support grants and £25 million in housing subsidy from central government during the period 1981-86 (*Doncaster Star*, 29 April 1987).

Even a non-Keynesian economist would find it difficult to argue convincingly that a rejuvenation of a local economy was a likely prospect where that economy had faced both a fall-off in public expenditure allocated to it, and simultaneously suffered deindustrialisation in its traditional economic sectors of coal and engineering.

It is important to acknowledge, also, that Task Forces did not bring with them into inner-urban policy any 'new' or additional resources. All second-phase Task Force money came from existing government departmental budgets. What the Task Force policy did was to concentrate monies already accounted for into inner-urban areas considered to be in greatest need (Platt and Lewis, 1988; Lawless, 1989).

An abiding and central feature of the Thatcher government's inner-urban policies was, as Murrey Stewart (1987) had it:

'an elevation of the role of the private sector both in the philosophy and the practice of policy'.

A constant theme under the Thatcher administrations was a belief that the private sector should be encouraged to invest in inner-urban localities, and that this would provide the basis for regeneration and renewal. This was a view shared by Michael Heseltine in his temporary role of 'Minister for Merseyside' (Parkinson and Duffy, 1984).

And so the term 'leverage' became more common in the urban policy lexicon, meaning the spending of some public money in order to lever, if successful, a lot more out of the private sector. This had been important under the Labour government of 1974-79, but under the Thatcher administrations was accorded a major priority.

Success on leverage seemed to have been limited in this case, however. Doncaster Task Force estimated in November 1990 that the total value of the 119 projects it had been involved in was around £16 million. The bulk of this finance had come from the taxpayer, either through the Task Force's own budget, or through the use of money already earmarked from Employment Training or the Youth Training Scheme. The Task Force estimated that just over £1 million of the £16 million had been 'levered' out of the private sector. It conceded, however, that most of this was not made up of monetary contributions but was, rather, the value of the time and services given by people from the private sector. An example would be the secondment by ICI of a production manager for one year to take charge of the attempt to rejuvenate the local Chamber of Commerce and Industry, detailed as an objective in the Task Force's action plan of 1988. ICI paid his salary for the year. Time given by Asda staff to Task Force projects, (dealt with later), would be another example.

Thus the economic regeneration strategy pursued by Doncaster Task Force was almost wholly a publicly-funded exercise. The objective of involving the private sector in the economic regeneration process was achieved in Doncaster then only in a more limited sense. It was limited to the private sector being involved in the **implementation** of policy decisions, such as the attempt to rejuvenate the Chamber of Commerce, or in the development of training packages which, without stretching the term too far, could be seen as involvement in at least the detail of policy design.

7.6 Doncaster Chamber of Commerce and Industry

In an attempt to assess the effectiveness of an economic regeneration scheme such as the one pursued by Doncaster Task Force, it is worth examining specific projects that were pursued. One such was the attempt to rejuvenate the Doncaster Chamber of Commerce and Industry. Following Task Force assistance, this process commenced in earnest in December 1990.

Again, the policy thrust here reflected an attempt at a public-private sector joint initiative. The idea of rejuvenation came from the Chamber itself, which had been in decline for some time and feared disappearing altogether.

The Task Force was approached and provided a one-off grant of £18,500. From the private sector came the idea, and the one-year secondment of a production manager from ICI, and from the public sector the grant. Essentially, the £18,500 was used to buy some computing equipment, and to launch a recruitment strategy with more professional-looking stationery, which formed the starting point of the 'rejuvenation'.

The theme of helping the small business sector as a central thrust of economic regeneration policy re-emerged here. The Chamber believed that the big and medium-sized companies had the resources to look after themselves. In its new, post-Task Force phase, its services were to be geared to the small business sector.

It was of some importance, moreover, that the process of attempting to rejuvenate the Chamber of Commerce had commenced **prior** to Task Force involvement. In 1987, the Chamber started to become involved in central government-funded training initiatives, and was increasing its income through those means. That the Chamber 'rejuvenation' was underway prior to Task Force involvement was also evident from membership figures. It had about 100 members in 1985, compared to about 300 in November 1990, before its Task Force funded recruitment strategy had commenced.

The target of the Chamber was to increase its membership to 1,500. The increase in membership fees that this would bring would ensure its survival, as the Task Force grant itself would not.

In the post-Task Force twelve months to November 1991, the membership recruitment strategy was showing some success: membership had increased to 400, a faster rate of growth than in the pre-Task Force era. That might indicate some modest benefit to the Chamber from Task Force involvement.

Essentially, the 'rejuvenated' Chamber of Commerce and Industry intended to apply itself to promoting small business growth and the growth of the small business sector. Its mechanism for so doing was an improvement in the quality and quantity of the services provided to that sector, such as health and safety auditing, organising training packages for companies, provision of desk top publishing facilities, through to finding an interpreter for the proprietor of a Chinese restaurant wishing to draw up Anglicised versions of the menu. A major part of its strategy was to provide 'business seminars.' where people from big companies in the area like British Rail, ICI, Case, Yorkshire Electricity, made an appearance to offer guidance and counselling to representatives of the small business sector.

What could be said, then, of this part of the Task Force's activities? On the positive side, any endeavour to stimulate a small business sector in a relatively depressed local economy was a worthy activity. And, also, this one was relatively cheap, costing the taxpayer a mere £18,500, not enough to pay an executive's salary. On the less positive, it appeared as if a rejuvenation strategy at the local Chamber of Commerce and Industry was underway anyway. And, whilst it benefited from the Task Force grant, the Chamber's view was that it had not needed any Task Force prompting to start the rejuvenation project. A realistic appraisal of the Task Force objective of rejuvenating the Chamber had to be that, even assuming the objective was achieved, the effect on the local economy could be only peripheral. The provision of business advice to people running corner shops and Chinese restaurants, though worthy in itself, was not going to provide sufficient economic regeneration impetus to overcome problems caused by massive deindustrialisation in coal and engineering.

7.7 Training

A major focus of the Doncaster Task Force was on training. The Task Force itself put forward the belief that it was engaged in a 'training-led regeneration'. They put together what they called 'customised training': discovering what the requirements of local employers were, and then meeting those requirements.

Numerous 'customised training' projects were organised. Asda, for example, located a new supermarket in the area in November 1990. Training courses were organised by the Task

Force for 200 residents of the Task Force area, after consultation with Asda management as to what these should consist of. In practice, the courses started off on confidence-building, and on overcoming return-to-work problems, for those who had been absent from work for some time, before turning to retail-specific skills.

Asda's commitment was to guarantee each Task Force trainee an interview. The success rate for this positive-discrimination measure was relatively high: 128 of the 200 trainees, or 64 per cent, obtained long-term jobs with Asda. The Task Force believed that many of the trainees who had not found work with Asda would eventually find it with other supermarket employers in the vicinity.

A measure of the economic plight of the locality was reflected in the fact that over 4,000 applications were received for the 300 jobs Asda had to offer (*Doncaster Star*, 15 August 1990).

Task Force committed £49,318 towards this £131,000 Employment Training package for Asda; and justified this expenditure, and this strategy, by saying that those it organised the training of for Asda would probably not have obtained jobs, or even interviews, with that company without the training. Asda's input into the programme was not financial, but was rather a commitment of staff time.

There are two aspects of this policy approach which might invite critical comment, however. Firstly, the policy constitutes a form of positive discrimination where, in terms of this immediate locality, it would be difficult to justify it. It was shown that both the Task Force and the local authority believed there were localities outside the Task Force territory that were equally, or even more, economically deprived than those within.

Secondly, it could be argued that providing public funds to a private sector company like Asda to enable it to engage in training was effectively a government subsidy to the private sector. Elsewhere, the Conservative administration first elected in 1979 frowned upon subsidies to business and industry. Surely

179

— although it was on a larger scale the principle is the same — it was the British nationalised coal industry's constant draw on government subsidies that induced its major retrenchment under the Conservatives in the 1980s and early 1990s. In any case, did Asda, with its £225.1 million operating profit in 1990 (Asda Group plc, 1990), really need taxpayers' money in order to carry out its labour training? The riposte would be to say no, of course it did not, but in the absence of that programme, the women returning to work, and the long-term unemployed who benefited from the Task Force's training project (*Doncaster Task Force*, 1990), might otherwise have not found jobs.

There were many other examples of training schemes organised by Doncaster Task Force, ranging from computer programming courses, through to bakery, bricklaying and warehousing skills, through to desk-top publishing and reprographics for disabled people (*Doncaster Task Force*, 1990). Such a commitment to improving the chances of gaining employment for people was commendable, though it remained a pity that the beneficiaries of such schemes were limited to those coming from a particular territory.

One interesting example of a training scheme organised by Doncaster Task Force related to Craven Tasker, a manufacturer of trailers for articulated lorries.

Craven Tasker moved its operations from Sheffield to within the Doncaster Task Force area in August 1990. The Task Force organised a training package for 30 residents of its area, again on the basis that Craven Tasker would offer each of the trainees an interview. As this was an 'economic activity transference', however, it deserved some examination.

The first point was that Craven Tasker, as a company, stood to benefit from moving into their new premises at Doncaster, and away from premises at Sheffield, which were 'old and not designed to facilitate modern manufacturing methods' (Craven Tasker, 1990). In that sense Craven Tasker's move would have a mildly positive effect on the South Yorkshire economy

180

as a whole, as the operating conditions of a medium-sized company were slightly enhanced. Secondly, Craven Tasker reported that 90 per cent of its new workforce had been recruited 'locally'; therefore the economy of the Task Force locality, and the people in it, also benefited.

The potential criticism that arose, however, was that the effect of Craven Tasker's move from Sheffield to Doncaster could be only marginal in relation to the wider South Yorkshire economy. Sheffield is all of 15 miles from Doncaster . This stricture on economic activity transference was expressed in an OECD (1987) review of urban economic policy. The argument of the OECD was that urban economic policy in these circumstances failed to add to overall economic performance, and merely shifted resources from one part of a national economy to another. Craven Tasker gave all their 160 employees in Sheffield a chance to work at the Doncaster site, though many declined, apparently due to transport difficulties (*Doncaster Star*, 7 September 1990). Craven Tasker, at the time this research was carried out, employed a similar number of workers at their Doncaster operation as they had at Sheffield. Economic activity transference here, then, benefited from a central government subsidised labour training programme. In these circumstances, the most positive way of viewing the activities of the Doncaster Task Force was that they were engaged in helping, marginally, to 'modernise' the South Yorkshire economy. To be viewed as having contributed anything more beneficial than that to the South Yorkshire economy would have required either, firstly, an increase in employment levels at Craven Tasker as a consequence of Task Force activity or, secondly, evidence that Task Force activity was an incentive for the initial relocation. That would have proved that there was a Task Force influence on decision-making by the management of Craven Tasker. Craven Tasker reported, however, that:

'We would have moved to Doncaster

irrespective of whether there had been any task force. It did not influence our decision at all' (Craven Tasker, 1990).

Similarly, Asda were also already on their way to establishing a supermarket in the locality under discussion. That is not to argue that the Task Force were claiming any credit for the relocations, simply to make the point clear to the reader that the Task Force itself was not influential in relation to the management decisions of Asda and Craven Tasker on location and relocation.

If the Doncaster Task Force's economic regeneration strategy was based on training, then from the point of view of an individual seeking economic advancement and the new career opportunities, that was potentially positive. And, still on the positive side, it could be argued that the Task Force's emphasis on training helped to bridge, in this particular locality, what was perceived by Haughton and Roberts (1990) as a 'gap' in central government urban policy. They argued, generally, that this 'gap' was a failure in the mid and late-1990s 'to plan for labour market expansion through training schemes.' Where a local economy continued to be depressed over a period, however, the implementation of training projects also begs the question: training for what? The acid test of whether a regeneration strategy such as this 'worked' or not ultimately rested on the effect it had on the unemployment levels within the territory under the jurisdiction of the Task Force.

Here the Task Force did not fare as well as presumably it would have hoped to. Whilst unemployment rates declined in the Task Force area when 1990 is viewed in relation to 1986, none of the relative decreases were as large as the relative decrease for Doncaster Metropolitan Borough as a whole.

Unemployment in Doncaster as a whole stood at 14.1 per cent in 1988, and 8.8 per cent in May 1990: a relative fall of 38 per cent. In the Task force area of Bentley Central it stood at 18.6 per cent in 1988, and had fallen to 12.9 per cent by May 1990: a relative fall of 30.5

per cent. In another Task Force area, Doncaster Central, unemployment stood at 19.8 per cent in 1988, and had fallen to 13.4 per cent by May 1990: a relative fall of 32.4 per cent (Doncaster Metropolitan Borough Council, 1987, 1989 and 1990).

7.8 Conclusions

It is easy, perhaps, too easy, to be negative about economic regeneration schemes such as the one examined here. There is no way of knowing, for example, what would have happened in the absence of the Task Force: perhaps that part of Doncaster under its jurisdiction would have fared worse than it did economically. Perhaps, although it seems unlikely, the Task Force contributed to an economic foundation for a regeneration yet to come.

Additionally, with its limited budget and powers, it is difficult to see what more the Task Force in Doncaster could have done. It did not have the resources, nor the time, in its short life span, to even attempt to influence the **industrial structure** of Doncaster. In that sense, what happened here was a mis-match of policy to problems, or at best a **partial** response to difficulties, for clearly Doncaster's economic problems were the result of an enforced change of industrial structure beyond its control: pit closures and retrenchment in other sectors of traditional industry.

In its own terms, the Task Force can lay claim to having gone some way towards achieving at least one of the four 'E's', or central objectives. It **did** appear to have improved the 'employability', through its customised training activities, of some of those within its boundaries. But, as noted earlier, there are, especially in this particular case, some moral problems associated with this kind of positive discrimination, if others similarly disadvantaged are not helped also.

The achievement of the objective of improving the environment was marginal, and this

183

was effectively recognised by the Task Force in its later action plans.

As to the achievement of the objective of encouraging 'enterprise', this was difficult to assess because 'enterprise' was not clearly defined. Training, and financial and managerial assistance, was provided or organised by the Task Force, and in theory recipients could have built on this to engage in 'enterprise'. If 'enterprise' was something to do with turning mere mortals into entrepreneurs, however, there was little evidence to suggest that this did occur, or even whether it was within the realms of policy possibilities.

In the terms of the final objective, the encouragement of more employment in the Task Force locality, there was some success but it was tempered by the fact that, immediately outside the Task Force area, there was more success, in relative terms.

Nor was the Task Force particularly successful in levering money out of the private sector. Indeed, far more money went the other way: from the public to the private sector.

At best, all that a small Task Force team, on a small budget, could really be expected to achieve in a locality such as Doncaster, which was facing rapid deindustrialisation, was partial and marginal. Within that constraint, the Doncaster Task Force could claim at least modest success.

The Conservative central government, led by Margaret Thatcher when the Task Forces were first established, and subsequently John Major, continued to believe in Task Forces in 1990. Perhaps, as suggested earlier, this was because like other measures under the Conservatives, it enabled central government to by-pass local government and take centralised control of specifically targeted urban policy. Whatever the reasons, the Conservative government's continued faith in Task Forces was demonstrated by the fact that the closure of Doncaster Task Force was to enable resources to be made available for a new Task Force, this time in Derby.

8 Barnsley Business and Innovation Centre: The impact of a strategy for innovation

8.1 The idea

The innovation centre idea started off in the early 1970s in the United States of America, where the first innovation centres were associated with universities (Leigh and North, 1986). In Britain, the first innovation centre, Strathclyde Innovation, was established in 1986 (*Northern Flagship*, undated), closely followed by the establishment of Barnsley Business and Innovation Centre (BBIC), where the actual centre was physically constructed by January 1988, and the first 'innovator/entrepreneur' moved in June 1988. As of March 1992, it was one of nine innovation centres operating within the United Kingdom (see Table 6) that were also members of the European Business and Innovation Centre Network (see later). The Barnsley Business and Innovation Centre was the only innovation centre to be operating during this time period in the heart of what had become a rapidly declining coalfield.

This study is based on a face-to-face

interview with the chief executive of the Barnsley Business and Innovation Centre, and interviews with 13 owner /managers of companies representing 17 businesses. At the time of the interviews, there were a total of 16 owner/managers associated with 20 businesses 'registered' within the BBIC. In other words, some owner/managers owned more than one business name within the BBIC. All the interviews took place in March and April 1992.

The idea behind innovation centres was straightforward: innovative products, processes or services could be turned into 'market winners', thereby stimulating both local and national economic development and aiding economic modernisation (see, for example, Hall, 1986; Rothwell, 1986). Business and innovation centres existed in order to help companies or individuals who might not otherwise be in circumstances propitious to the development of an innovation (*Northern Flagship*, undated). Improving the innovative capacity of, and within, a company, it was argued, would improve the competitiveness of the economy as a whole. In particular, this would be achieved by the encouragement and fostering of the development of an idea into a marketable product, process or service:

'Successful commercialisation of patented innovations will lead to the formation of new businesses or extend the life of existing businesses (by opening up new markets)' (Leigh and North, 1986).

Similarly, a network of innovation centres, if successful, could work towards preventing a monopolisation of the 'agenda for innovation' by established large firms. It has been argued that the latter cannot be relied upon as a source of technical progress (Stroetmann, 1979); and that case could be made particularly strongly where an innovation made entry into market provision of a product, process or service more easy for another company or companies with limited resources. In such circumstances, an established large company

Table 6

UK Business and Innovation Centres that were also Members of
the European Business and Innovation Centre Network as of
March 1992

Name	Address
Barnsley Business and Innovation Centre	Innovation Way, Barnsley, S71 1JL
Birmingham Technology Ltd	Love Lane, Aston Triangle, Birmingham, B7 4BJ
Cheshire Business and Innovation Centre	Innovation House, 6 Seymour Court, Tudor Road, Manor Park, Runcorn, WA7 1SY
Greater Manchester Business Innovation Centre	Windmill Lane, Denton, Greater Manchester, M34 3QS
Lancashire Business and Innovation Centre	Suite 302, Daiseyfield Business Centre, Appleby Street, Blackburn, BB1 3BL
Newtech Innovation Centre	Deeside Industrial Park, Deeside, Clywd, CH5 2NT
Northern Ireland Innovation Programme	Innovation Centre, NORIBIC, Springrowth House, Balliniska Road, Springtown Industrial Estate, Londonderry, N. Ireland, BT48 ONA
Innovation Wales	Cardiff Business Technology Centre, Senghennydd Road, Cardiff, CF2 4AY
Strathclyde Innovation	Unit A1, Building 1, Templeton Business Centre, 62 Templeton Street, Glasgow, G40 1DA

Source: *Northern Flagship* (undated, circa 1992)

might be tempted to suppress innovation. It has been argued elsewhere that the British national economy in the 1970s and 1980s was a 'concentrated' economy where, at least in the manufacturing sector 'big business' was dominant (see Grant and Nath, 1984, p.5-63). Williams et al (1983) argued the British economy was of such a concentration that:

> 'The trade performance of British manufacturing is largely the result of the business strategy and the market place success or failure of 100 or 200 giant firms.'

Presumably, the same could be argued in relation to innovation. If, in this 'concentrated' economy, the strategies for innovation of the large established companies were inefficient or ineffective, that in turn would have a deleterious effect on the national economy as a whole. If a network of business and innovation centres could have an — admittedly at best marginal — impact on reversing the potential problem, then a marginal impact was better than none at all.

Moreover, in relation to coal mining localities, it was noted earlier that Hall (1988) had argued that 'new' industries, with potential for growth — such as robotics, biotechnology, 'alternative' energy systems — would not be likely to locate in territories associated with traditional industry. If this was the case, then clearly there was an impetus for intervention by governmental, quasi-governmental or other regenerative organisations to ensure that 'new' industrial and commercial sector **did** establish themselves in localities formerly associated with traditional industry. Otherwise, those localities would fail to modernise, would remain dependent upon a declining traditional industrial base, and become progressively less prosperous than other local economies within the national economy. The establishment of a business and innovation centre might be an attempt, at least, to modernise a local economy in these circumstances.

An economic regenerative strategy which focused on innovation and modernisation was bound to command a measure of across-the-board political support. It was, of course, theoretically possible that those strongly committed to the principles of the free market could be against the establishment and maintenance of business and innovation centres. The establishment of business and innovation centres might constitute, from their perspective, an unwarranted intervention in the market. Innovators would innovate without government or quasi-government intervention, and if their products, new processes or services were worthy, would always be able to find the appropriate finance. Burton (1983), for example, argued that:

> 'Neither standard economic theory nor everyday experience offers any ground for the belief that politicians and bureaucrats are more alert in "picking winners" of the future than private entrepreneurs motivated by opportunities for personal profit and possessed of specialised knowledge of business methods and markets.'

But, aside from those who would wish to adhere rigidly to free market principles, elsewhere on the right in politics, policies designed to stimulate innovation were likely to receive support. Peter Lilley, for example, normally associated with the free market wing of the Conservative Party, established, in 1991 as Secretary of State for Trade and Industry, an 'Innovation Unit' at the Department of Trade and Industry. This involved the secondment of five representatives of business from the private sector to the DTI, who were each to specialise in a particular activity, working with the regional offices of the DTI. These activities might include the encouragement of the provision of 'venture capital for small firms, improving the commercial exploitation of the UK's science and technology base and developing innovation teaching modules for business schools' (*In*

Business Now, 1992)

Lilley had apparently identified 'the poor status of innovation — the successful exploitation of new ideas — as a major obstacle to Britain's industrial success' (*In Business Now,* 1992).

Labour Party support for strategies to stimulate innovation was also evident in the 1980s and early 1990s. Firstly, from their more favourable disposition generally to intervene in the industrial economy (Turner, 1989). Secondly, at the local level, from the financial and other support given to Barnsley Business and Innovation Centre. Barnsley's continually solid Labour council (Turner, 1988) provided a £100,000 grant for the centre at the start of the operations, and provided the land upon which the centre was situated. Lord Mason of Barnsley, formerly Roy Mason, who had been in the Labour cabinets of the 1960s and 1970s variously as Postmaster General, Minister of Power, President of the Board of Trade, and Secretary of State for Northern Ireland, was chairman of the BBIC at the time of this study. That chairmanship provided another symbolic example of the Labour Party's support for this, and by extension other, strategies for the stimulation of innovation.

8.2 Technological modernisation and government intervention

Government, or government-agency, intervention in the economy to try to secure technological modernisation, and by that mechanism achieve a competitive advantage for the economy, has a long history. The established of business and innovation centres — albeit at the instigation of the CEC rather than national government — can be seen as being part of that history. And because the end objective of technological modernisation was economic regeneration, a study of an innovation centre, and a mention of the longer tradition of efforts to ensure this king of advancement, could legitimately be located within a work such as this.

The 'history' of government intervention to support technological modernisation is too extensive to document and analyse in full here. Nevertheless, the establishment and work of certain institutions can be mentioned in order to provide an appreciation of that history. The National Research and Development Corporation, for example, which was merged with the vastly slimmed down National Enterprise Board in the early 1980s to form the British Technology Group, began granting financial support for technological innovation and the exploitation of inventions in 1949. On its election in 1964 after 13 years of Conservative rule, Harold Wilson's Labour government established the interventionist Ministry of Technology ((or 'Mintech', as it became known). Frank Cousins was its first Minister, from October 1964 to July 1966, when Anthony Wedgewood Benn took over. The then developing computer industry in Britain was one of the sectors it aided with public money.

The 1970s also saw government schemes to support and encourage the high-technology sector. The Product and Process Development Scheme, for example, was launched in 1977 to accelerate product and process innovations through a subsidy to firms' development costs. The Microprocessor Application Project, launched in 1978, was another Labour government scheme aimed at encouraging microelectronic applications in industry, by retraining of engineers and technicians, providing grants for new technology, and publishing its benefits (Grant, 1982).

The impetus for government intervention in this business sector was obvious: Britain's industrial and economic future must lie in what it is or should be capable of doing better than companies in competing economies. From the perspective of the early 1990s, this might mean further expansion of producer service industries: education, tourism, insurance, pensions and other financial services. In strictly industrial terms, it would also have meant expanding the high-technology industries, and 'high tech' applications within other traditional ones. The continuation of traditional work practices and processes, and traditional industries, would see companies in the British economy increasingly trying to compete against companies in countries where labour costs were far lower, or against companies which had introduced high-technology on a large scale.

'Technological modernisation', the term 'high-tech', and connected phraseology, were, of course, catch-all descriptions, used to refer to a multitude of industries, industrial sectors, products and processes, or areas for research. Robotics, fibre optics, microelectronics, information technology, space research and the space industry, the development of new and improved materials in areas such as plastics, ceramics, special steels and other alloys, were all examples of where 'technological modernisation' had been at work (Turner, 1989).

It was axiomatic that advanced industrial economies which did not achieve technological modernisation successfully would find competition increasingly difficult against companies and industries abroad that had achieved technological modernisation successfully. There was some evidence to suggest that in the 1980s the British economy was not achieving technological modernisation at a sufficiently rapid rate. It was argued, for example, that companies in the British economy were employing the microchip in factories, for example, less extensively than companies in the economies not just of the USA, Japan and West Germany, but also of smaller nations such as Sweden and Denmark (Northcott with Walling, 1988).

An argument in favour of selective government assistance to the new technology areas was that for some of them the capital investment would have been so high, and the returns from that investment such a long way off, that only multinational enterprises would have been able to make the necessary investment. In such circumstances, a country might find itself without a domestic base in a major new industrial sector, if the multinationals located their research and development elsewhere. For those who adhered to a free market approach to the economy, this might not have been a problem: it would simply be a reflection of companies within certain national economies specialising in what they do best. And, according to the economists' 'law' of comparative advantage, this would work through to the eventual benefit of all.

For others, however, an absence of a domestic base in a major new industrial sector **was** perceived as a problem. It was this kind of perception which led the Labour government of 1974-9 to support the 'transputer' company Inmos under the aegis of the National Enterprise Board, for example.

Indeed, support for the high-tech sector was one of the areas where the differences between Labour and Conservative governments have frequently been differences of degree and emphasis, rather than of substance. It was Harold Wilson, for example, who managed to hitch science to socialism and talk about forging a future in the 'white heat of the technological revolution' in the 1960s. And despite its free-market stance, and opposition to active government intervention in the economy, the Conservative party's 1987 general election manifesto positively boasted about the government's support for research and development:

> 'Government support for research and development amounts to more than £4.5 billion per year. It is larger as a share of our national income than that of the United States, Japan or Germany. A country of our size cannot afford to do everything ... The task of government is to support basic research and to contribute where business cannot realistically be expected to carry all the risks' (Reprinted from *The Times Guide to the House of Commons*, June 1987).

For the Labour Party, however, support for high-technology industry under the Thatcher government was not enough. Roy Hattersley, Deputy Leader of the party, argued in 1985 that:

> 'The Tories have become the latter day Luddites with their emphasis on industry that is not so much low tech as no tech. Their priority is low pay and the substitution of workers for machines' (*The Guardian*, 6 June 1985).

Support for innovation and technological modernisation continued under the Conservative central governments of 1979 through to the early 1990s, despite its general predilection against government intervention at the level of the firm in the economy. Much of the government support was directed at small and medium sized companies; some was directed at wider schemes of technological research.

In 1988 there was the launch of the Link programme, for example. A series of projects lasting up to six years, with around £420 million in funding from government and industry, Link sought to forge collaboration between academic, government and industrial laboratories. The first five Link schemes, announced in February 1988, covered molecular electronics, advanced semi-conductor materials, industrial measurement systems, eukaryotic genetic engineering and nanotechnology. They were intended to be 'pre-competitive' research projects, where companies involved would be free to go their own way using the technology once the research programme finished. The view of Lord Young, Secretary of State for Trade and Industry at the time, was that the idea was to 'encourage industry to do something that is fairly unnatural for industry — to work with educational establishments' (*Financial Times*, 4 February 1988).

The Conservative government of this period also attempted to strengthen the information technology industry in Britain, by the £350 million Alvey research programme. This was a five-year programme, introduced in 1983, and was again an attempt to marshall the resources of industry, government and the universities. One of the projects it worked on was the fifth-generation computer. Alvey was launched, in large part, as a response to widespread fears that the Japanese and US information technology industries were leaving Britain's at the starting grid.

Many of the other schemes to support technological modernisation in the 1980s and early 1990s were directed at individual small or medium-sized firms.

Regional Innovation Grants, for example, were available from the Department of Trade and Industry to firms of less than 25 employees in certain specified localities designated Development Areas under regional policy. These grants were designed to finance 50 per cent of the development of an innovative project up to a maximum grant level of £25,000 (*Northern Flagship*, undated, circa 1992). At least 3 companies in Barnsley Business and Innovation Centre had received Regional Innovation Grants as of March/April 1992.

SMART — Small Firms Merit Award for Research and Technology — was another DTI sponsored scheme in the 1980s and 1990s for helping firms with under 50 employees involved, or seeking to become involved, in technological innovation. It was a competition divided into two stages. Winners of Stage 1 of the competition received £45,000 towards project development costs and the right to enter Stage II of the competition. Winners of Stage II would receive finance of up to £60,000, as of 1992, to develop their projects (*Northern Flagship*, undated, circa 1992). One of the functions of Barnsley Business and Innovation Centre was to help small companies located inside their premises prepare for competitions such as SMART. In this, some success was evident at BBIC. Dena Technology, a company within the BBIC premises pioneering innovative techniques of ceramic mixing, was a winner of both SMART 1 and SMART 2 awards; it was one of the only eight firms from Yorkshire and Humberside to progress to the finals of the SMART 2 awards (BBIC Autumn, 1991). Dena

Technology had also benefited from seedcorn funding from the BBIC itself, where royalties would be payable to BBIC were Dena Technology's services to become successful commercially.

There were a collection of other activities by central government in the late 1980s and early 1990s aimed at the stimulation of technological modernisation and innovation at the level of the firm, especially the small firm. SPUR, for example, was a scheme to assist firms with up to 500 employees develop new processes or products, where that would 'enhance technology levels in the industry as a whole' (*Northern Flagship*, undated, circa 1992). The Department of Trade and Industry White Paper, *DTI — The department for Enterprise*, published in January 1988, also announced new measures to assist small and medium sized companies which, if not tied exclusively to innovation, were certainly aimed at improving design, marketing, quality management and manufacturing systems. The project was that government would help businesses with under 500 employees seeking to improve in these directions, by supporting these companies financially to draw on the services of consultants. A fund of £50 million was provided for this purpose for 1988-9, initially supporting 1,000 consultancy projects a month. Half the cost of the consultancy project would be met by the company; the other half by the government. In depressed regions, the government would fund two-thirds of the cost (DTI, 1988).

Of significance in relation to government and its input into technological modernisation efforts, however, was that the January 1988 White Paper announced the end of most grants to individual firms for high-technology projects. It did, nevertheless, state that a few exceptions would be made, and that high-technology research would continue under the European Community's 'Esprit' project (DTI, 1988).

In fact, there were a number of schemes announced by the CEC in the 1980s aimed at the technological modernisation of business and industry across the EC member states. This

196

emphasised that government intervention in the economy could come at a number of different levels: local government level; national government level; international government level (EC).

March 1985, for example, saw the CEC announce Brite (Basic Research in Industrial Technologies for Europe). Brite was a transnational programme involving 432 firms, universities and research institutes across all of the twelve member countries. It had Community funding of £80 million over 4 years, and was aimed at projects designed to help the Community's traditional industries, such as chemicals and car manufacture, become more competitive with those of Japan and the USA. It followed the model of a joint public-private initiative, with the consortia involved expected to match the £80 million government funding. Brite was granted more European money for a second phase to run from 1988 to 1991 (Turner, 1989).

ESPRIT was approved by the European Community in February 1984. ESPRIT, the flagship of the Community's industrial research programme, was a £471 million five-year project, again bringing together companies and research institutes across the Community, in an effort to create a competitive European capacity in information technology. ESPRIT was seen by its supporters as having the role of acting as a catalyst to pre-competitive research by companies in different Community countries. Half the cost of research initiated in this way was paid by the Community, the other half by the companies themselves. Such research might not otherwise have taken place according to advocates of ESPRIT, without this governmental 'spur'. Moreover, they say, projects such as these avoid duplicating effort in the same area by different companies, and give a European scale to research. A second phase of ESPRIT, with European funding of £1.1 billion, was announced in December 1987.

A further project, launched at the end of 1985, which had eighteen participating European countries including six outside the EC, was

Eureka. Eureka was the brainchild of the French, who were at the time committed to the state-funding of projects, and was also aimed at the high-tech sector. Again, the focus of attention was the need to compete with companies in the US and Japan. The belief of the advocates of this project was that co-operation could be achieved across the whole of Western Europe, stretching beyond research into marketing collaboration and, in the meantime, avoiding the delaying politics and bureaucracy of the European Community. The first ten projects announced in November 1985 under the aegis of Eureka were concerned with microcomputers, compact vector computers, solar power, robotic lasers, membrane microfilters, high performance cutting and welding lasers, seeing robots, medical diagnostic kits, data collection on airborne pollution, and computer network research (Turner, 1989)

Examples of national and international governmental efforts to stimulate innovation and technological modernisation could be seen then to have taken a variety of forms, some of which have been discussed above. There were also cases of attempts to stimulate technological modernisation at the **local** level. An example of a local level intervention to secure the same might be science parks. Here, from about the mid-1980s onwards, local authorities and local development agencies began to collaborate with academic institutions in developing business parks where the emphasis was on knowledge-based business, and on technological modernisation. Like innovation centres, the science park idea was another one which started in the USA. Science parks had been established there since the 1950s. They first spread to Britain in 1971, with science parks being established at Heriot-Watt and Cambridge Universities. By 1990, there were 39 'true' science parks in Britain (*Northern Flagship*, undated, circa 1992).

The idea behind innovation centres, then, could be seen as fitting into a consensus on the need for, or desirability of, some degree of government support for technological modernisation in firms. It was a consensus which appeared to stretch across different levels of government, and across the ideological spectrum.

But with innovation centres, the method of policy implementation was different to the other policies aimed at technological modernisation referred to. To start with, they were territorially-specific: their services were, in theory, available to all but in practice most clients would be from within the vicinity, or would have to move to a particular vicinity to partake of the services. This contrasted with schemes operated by, say, the DTI, which would be open to any qualifying firm across the country. Secondly, the idea of innovation here — in line with some DTI schemes, as noted above, but out of line with 'grander' schemes for technological modernisation, such as the National Enterprise Board or ESPRIT — was conflated with entrepreneurship. The 'new entrepreneur' would also be an innovator, and his or her success in the latter would lead to success in the former.

8.3   The structure of innovation centres

Innovation centres were not uniformly-alike organisations. Leigh and North (1986), for example, identified five options for activity that may be pursued singularly or in combination with each other. These were:

(i)   'innovation counselling'. This entailed the giving of advice to individuals who had an innovative idea for a product, process or service. The advice would focus on the technical and commercial feasibility of the idea;

(ii)   'project appraisal and development'. Here, the innovation counselling is built on and the innovation centre becomes more closely involved with the development of the innovation. This might entail an involvement by the innovation centre in:

a. prototype design, manufacture and testing to production stages, and

b. establishing ownership rights over the innovation by patent protection or design registration' (Leigh and North, 1986);

(iii) 'project appraisal and development with workshop facilities'. Same as above with the 'internalisation' of the development of products and services: in other words, these were carried out by innovation centre clients on the premises;

(iv) 'management of innovation process and enterprise formation'. Here the staff of the innovation centre would have a much closer involvement 'in the detailed management of the innovation-development process'. The staff would help in a much more involved way in the development of the innovation and were much more heavily involved in the 'successful translation of projects into commercial production' (Leigh and North, 1986).

(v) 'technology transfer'. The objective here was that firms outside the innovation centre could have made available to them a 'bank' of innovative products and services, held in the innovation centre, upon which they could draw. The aim was to extend the advantages of economic specialisation in what was effectively a development of the law of comparative advantage. In other words, an established company outside of the innovation centre might find it cost effective, and a more efficient use of resources, to 'buy in' designs for innovative products and services from those who specialised in innovation: individual inventors; universities; other firms — domestic or foreign — who would make their innovations available on a royalty-payment basis.

8.4 Barnsley Business and Innovation Centre

The Barnsley Business and Innovation centre was probably best characterised as being in line with option (iii) above. It appraised the commercial and technical feasibility of projects before taking them on. Of the 20 to 25 enquiries received every month at Barnsley BIC, there were apparently only two or three that were considered to be of sufficient quality to take seriously: the rest of the enquiring parties were given counselling as to how their ideas might be developed, and referred to other enterprise agencies. Apparently, the varying depths of the national economic recession of the late 1980s and early 1990s made little difference to the number of enquiries received by the BBIC in any time period.

For those ideas that were taken on board, however, BBIC then divided its work into two. The first stage was the 'project stage', where clients were given help in developing their ideas, advised on where to look for finance, or advised as to how a project might be marketed. This input by BBIC was free of charge to clients, though if they were in the BBIC workshops they were still liable for rental payments. The second phase of the input by BBIC was where staff worked with clients on the development of a 3 year business plan. Where this involved work for BBIC staff — in, for example, designing mail shots, engaging in marketing, or putting the client company's accounts on the central computer — then a fee was agreed.

The option (iii) conditions were satisfied by BBIC's workshop provision — BBIC had 37 workshops in 40,000 square feet of space as of March 1992 — and an appreciation of the option (v) approach was also encompassed, if not fully developed. For, like most other UK innovation centres, Barnsley was part of a European-wide linked network of innovation centres called the European Business and Innovation Centre Network (EBN). This was a network instigated by the CEC. The objective was eventually to provide a database of product, process and service innovations that companies across the member states of the European Community could draw on; and to facilitate an interchange of those innovations between firms. Alternatively, products, services or processes developed in one innovation centre could be 'cloned': produced on the same basis in another innovation centre, somewhere within the European Community. Apparently, as of March 1992, the EBN was still at the development stage.

On a similar front, Barnsley Business and Innovation Centre was the home of the 'Technical Action Line' sponsored by the Department of Trade and Industry (BBIC, Autumn 1991). The objective here was to provide a rapid 'diagnostic service' to any company in the locality with a technical problem. These might range from trying to find which European Community standards applied to a company's products through to improving the quality of products. Initial enquiries were free, but if those operating the Technical Action Line felt it necessary to send a 'consultation diagnostician' or to recommend an 'expert consultant', for that there would be a fee to pay. In some cases, assistance towards these fees would be available from the Department of Trade and Industry. Apparently, the experts drawn upon for technical assistance included:

'... engineers, scientists, programmers, technicians and professionals from all sectors of industry, industrial support groups, R and D organisations and higher education institutes' (BBIC, undated, a)

Formally, the BBIC summarised the services it made available as including the following:

'free advice; modern premises; shared central services; project structuring;

business planning; grants and finance; project evaluation; seed capital; computerised systems; technical appraisal; technical development; production of prototypes; patenting and copyright; quality control; costing; sales; marketing; layouts; mailshots; brochures and public relations; sub-contracting; market research; market analysis; overseas sales; personnel; training; recruitment techniques; time management' (BBIC undated, b, circa 1991).

Amongst its staff of twelve, the BBIC had five key players: a chief executive with a PhD in engineering; an economics graduate and accountant employed as business development manager — part of his function was to help 'innovator-entrepreneurs' put together business plans and understand accounts; a marketing manager; a training manager; and a technology and innovation manager, who analysed the feasibility of the 'innovative' project being proposed to the BBIC.

Technical advice, then, was available in strength at the BBIC, as was workshop provision and central support services. In addition, the BBIC staff would make available their expertise in the search for venture capital funds and other sources of finance. Also available, for selected projects and clients, were small amounts of seed fund capital from the BBIC's own resources. These might be funds of £5,000, £10,000 or £15,000, intended to ensure the survival of a company through the development period of an 'innovation'. The provision of such funds by the BBIC was available on a loan/equity basis, where the funding provided would be up to a maximum of 20 per cent of the total equity in a recipient company. As well as providing a survival fund for companies, the long-term aspiration for such funded projects was to provide finance for the BBIC as client companies' projects achieved success. Hence the equity stake approach (BBIC, undated, c, circa 1991).

BBIC was also involved in the provision of

an 'industrial innovation' service for small companies called 'BBIC SWORD'. The way that this was operated at the BBIC was that local small businesses would each pay £600 to send a representative to attend workshops over a 40 week period. At the workshops those representatives would learn the art of 'generating ideas, selecting ideas, developing ideas and launching ideas'. The publicity material for the scheme made the slightly difficult to believe claim that someone attending the programme:

'will generate over 100 new product or service ideas that will build on your existing skills and strengths and also expand your product or service range' (BBIC, undated, d, circa 1991).

According to the publicity material:

'... SWORD has a 98 per cent success rate in firms adopting the system. And to date, around 1,200 companies worldwide have taken it on board, with one in 10 of the world's top organisations using SWORD to develop skills and the corporate culture which breeds innovation' (BBIC, Autumn 1991).

Ten 'fledgling firms' engaged in the SWORD programme in 1991, organised by the BBIC. Apparently, the idea of SWORD was to show companies how to engage in a 'low cost, low risk approach to developing new products or services to ensure growth without risky investment in plant and machinery, or diversification into unknown markets' (BBIC, undated, d, circa 1991). The BBIC was both a limited company and registered as an enterprise agency. One of its activities under its latter guise was the provision of advice in 1991 to 30 local companies on how best to approach their business activities in relation to the establishment of the Single European market in 1992/93. This was carried out in conjunction with the local Chamber of Commerce, and funded by the central government Department of Employment, with some private sector contributions under the DoE's Local Enterprise Agency Project Scheme (LEAPS) initiative (BBIC, Autumn 1991).

The BBIC also held a series of other seminars for local businesses in 1991 on themes ranging from understanding balance sheets, when and when not to use management consultants, and marketing in the Single European market (BBIC, Autumn 1991).

8.5 Locality and local economy

Literally across the road from Barnsley Business and Innovation Centre once stood Redbrook colliery. This had employed 476 prior to its closure at the end of 1987 (*Financial Times*, 17 October 1987). The locality in which the BBIC stood was at one time heavily-mined territory. Standing outside Barnsley Business and Innovation Centre in March 1992, the remnants of four collieries could still be seen in the near distance: Redbrook, North Gawber, Woolley and Royston Drift. Two of these were very modern pits: Royston Drift had opened in the 1970s; Redbrook, on the site of a much older 'satellite' shaft of the nearby Dodworth colliery, had not opened until after the 1984/85 coal industry dispute was over. Immediately prior to its opening, Redbrook colliery had seen investment of over £50 million directed towards it by British Coal (Redbrook/Woolley Campaign Group, undated). Within a five mile radius of Barnsley Business and Innovation Centre, 16 collieries, one collection of workshops and one administrative headquarters had closed since 1968 (see Table 7). The two collieries existing in 1968 that remained operational in 1992 — Grimethorpe and Houghton Main — saw substantial reductions in labour force totals amounting to partial closure. Grimethorpe and Houghton Main were amongst the 31 collieries announced for closure in October 1992. In addition the closure of Shafton workshops was announced in June 1993, ending all British Coal activity in the locality (*Barnsley Chronicle*, 11 June 1993). Barnsley was one of the only three **major** towns in England — the others being

205

Doncaster and Mansfield — where coal had such dominance in the 1970s and early 1980s. Clearly, then, this was a local economy where modernisation — either by spontaneous economic regeneration, or by a strategy of government or government agency intervention — was necessary in order to move it from a phase of reliance on traditional industry to one which provided a sustainable basis for economic development.

Also, obviously, on the economic base upon which the innovation strategy was being imposed, the operation would be a difficult one: the local economic 'culture' had for years been one of wage labour sold to traditional industry, rather than one of individuals 'innovating', or of people engaging in 'entrepreneurship'.

8.6 Budgets and agendas

As of March 1992, Barnsley Business and Innovation Centre was operating on a budget of around £500,000 per year. There were three public sector inputs into this: from the CEC; from Barnsley Metropolitan Borough Council; and from British Coal Enterprise.

The initiative behind innovation centres in general came from the CEC. In 1986 it informed all local authorities in the United Kingdom eligible for Urban Programme money from the UK central government of its project. Barnsley Metropolitan Borough Council launched a bid and won support at the European level in 1986.

Of the £500,000 budget referred to above, £197,000 came from the CEC in BBIC's first year, and £175,000 per year in the subsequent two years. Further funds from the Commission were being sought under the RECHAR scheme mentioned earlier: £130,000 being sought for 1992/93; £110,000 for 1993/94; £90,000 for 1994/95.

Another substantial public sector input of resources came from Barnsley Metropolitan Borough Council, which donated £100,000 at the start of BBIC's operations to cover a five-year period. It also provided, effectively free of charge, the six acres of land upon which the BBIC was situated.

The third public sector input came from

British Coal Enterprise, which provided a £50,000 grant at the start of BBIC's operations to cover a five-year period. The buildings which housed the BBIC all legally belonged to British Coal Enterprise but were 'gifted' on a peppercorn rent to BBIC on a 125 year lease.

Table 7

Collieries and associated works closed within a 5 mile radius of Barnsley Business and Innovation Centre

Colliery	Workforce	Workforce	Status April 1992
Barrow/Barnsley Main[1]	1164	624	Closed
Barrow Coke Works	350[2]	Closed	Closed
Birdwell (workshops)	213[3]	Closed	Closed
Darfield Main	1049	379	Closed
Dearne Valley	406	324	Closed
Dodworth/Redbrook[1]	1444	489	Closed
Ferrymoor/Riddings	438	382	Closed
Grimethorpe	1895	1279	909[4] workforce
Grimethorpe South Side Coal Preparation Plant	Not Known	Not Known	129[4] workforce
Grimethorpe Admin HQ	500[3]	Closed	Closed
Haig	606	Closed	Closed
Houghton Main	1572	1481	380[4]
Monk Bretton	913	Closed	Closed
North Gawber	1310	Closed	Closed
Rockingham	1550	Closed	Closed
Royston Drift	n/a	493	Closed
Shafton (workshops)	450[3]	Not Known	327 workforce
Wentworth Silkstone	530	Closed	Closed
Wharncliffe Woodmoor (4&5)	1121	Closed	Closed
Woolley	2041	838	Closed
Wombwell	890	Closed	Closed
Totals	**18442**	**6289**	**1745**

Source: *Guide to the Coalfields* (1968), Colliery Guardian, London; and Redbrook/Woolley Campaign Group (undated), *Keep Redbrook and Woolley Collieries Open*, Campaign Publicity Leaflet; *Coal News*, May 1991.

Notes to Table 7:

(1)   Barrow colliery workforce transferred to the Barnsley Main site in 1985; Dodworth colliery workforce transferred to the Redbrook site in the same year;

(2)   Figure here for 1975 from *The Times*, 1 November 1975;

(3)   Figure here is for workforce in 1983. Strong likelihood is that number employed in 1968 was higher;

(4)   Figures from British Coal April 1992.

Private sector financial and material
support came from a variety of organisations but
for the most part was marginal. The most
important private sector contribution was the
gift by British Telecom of a telephone system
for the first phase of the BBIC. Other private
sector organisations — ranging from local
utilities through to national high street
retailers, banks and local manufacturing
companies — donated small sums which went to
finance activities such as the publishing of
newsletters. Representative of these private
sector donations would be the £2,000 grant given
to the BBIC in February 1992 by the Whitbread
brewery (*Barnsley Chronicle*, 14 February 1992);
or the £1,000 received from Allied-Lyons in the
early 1990s, which formed part of a three year
sponsorship of the BBIC by that company (BBIC,
undated, circa 1991).

The deficit balance between what was
received in public sector funds and the £500,000
operating budget of the BBIC was bridged by,
chiefly, the monies collected in rent from
clients (small business operators) in the
workshops or incubator units at the BBIC. This
was supplemented by consultancy services and
other services for companies within and outside
the innovation centre.

Despite the emphasis placed, especially by
central government, on the desirability of
attracting private sector finance to
regeneration projects, the funding and
initiation of the Barnsley Business Innovation
Centre remained largely a public sector affair.

The evidence suggested, further, that there
was considerably more support for the Barnsley
Business and Innovation Centre emanating from
the CEC than there was for it from the British
central government of the time. The story of
how the BBIC was established provides some
evidence of this.

Before a business and innovation centre of
this type could be established, the Commission
insisted upon a pilot scheme. The objective of
a pilot scheme was to evaluate the chances of
success for a fully established innovation
centre. It would end with the writing of a five

year business plan: not the least of the
questions addressed within such a plan would be
how the innovation centre would be funded.

Under such a pilot scheme, the Commission
would agree to fund fifty per cent of an
innovation centre's costs, with the remainder
being financed from within the host country:
from local or central government funds, for
example, or a public sector regeneration agency.
This was, elsewhere in this work, discussed in
more length as the 'additionality' principle.

Such was the nature of central-local
government relations in the 1980s , that as the
decade progressed central government began to
take ever greater control of how local authority
budgets were spent (see Butcher et al, 1990, pp
68-72; Newton and Karran, 1985, pp 114-129;
Stoker, 1988, pp 151-172). Where monies being
spent by local authorities were directly
allocated by a central government department to
designated local authorities only — such as
with monies allocated under the Urban Programme
— this control was even tighter.   Urban
Programme money provided at least part of the
funds granted to Barnsley Business and
Innovation Centre by Barnsley Metropolitan
Borough Council.

Apparently, in this instance, the CEC
wanted the pilot scheme to last for 30 months:
the British central government would agree to
one year only.

This represented at least a partial
corroboration of the point stressed earlier (see
chapter two): that the issue of the 'need' to
regenerate the former coalfields found a greater
salience on the policy agenda of the CEC than it
did on the agenda of the Conservative central
government in Britain of 1979 to the early
1990s.

8.7 Objectives and criteria for assessment of projects

The notice board outside the Barnsley
Business and Innovation Centre outlined the
following as the objectives of BBIC:

'The purpose of the centre is to help the

growth of new and existing, small and
medium sized enterprises which have
innovative technology-based products,
processes or services'.

Broadly, the Chief Executive of BBIC
endorsed these views: the aim was to seek out
entrepreneurs with technically innovative
products and services and help them to achieve
their objectives.

What this led into, of course, was the
question of what exactly constitutes an
'innovative technology based product, process or
service'? The definition from economic science
was that innovation 'is usefully defined as
invention applied commercially for the first
time', and invention is 'the process whereby
conventional inputs and general knowledge are
combined to produce technical knowledge' (Rowley
1973, p33). Innovation and invention is defined
similarly elsewhere (see for example, Parker,
1978). In practice, however, at the Barnsley
Business and Innovation Centre, there appeared
here to be no hard and fast rules: discretion
was exercised by BBIC staff. They looked for a
degree of innovation in a project, and for
projects which had the potential for further
development. Semi-jocularly, the point was made
that if something 'had not already been done in
Barnsley, then it's innovative'. Essentially,
the point was to avoid clients who simply wanted
to replicate what was being done elsewhere.

Most studies of innovation within economics
focus on the role and impact of innovation at
the macro-economic level of the national
economy: management structures and the
development of innovation; multinationals and
innovation; the diffusion of innovation within
an economy; the 'competitive environment' and
innovation (see, for example, Hall, 1986;
Metcalfe, 1986; OECD 1971; Parker, 1978;
Rothwell, 1986; Williams 1986). This was not
the intention here, nor would it be particularly
useful or even possible to carry out such a
study relating to the activities of the BBIC.
Instead , this study was meant to be narrow and
specific, concentrating on the activities and

impact of one innovation centre over a short period of time.

The BBIC did not have a circumscribed catchment area. Providing their ideas were appropriate, anybody was welcome from any part of the UK — or for that matter, the globe — to locate in the BBIC. In practice, with the nearest alternative innovation centre about 30 miles away across the Pennines in Greater Manchester, the likely population the BBIC would draw on would live in Yorkshire, and potentially, north Derbyshire and north Nottinghamshire.

Leigh and North (1986) have argued that an innovation centre should have a catchment area of 'at least 3 million people', plus a university or polytechnic for the generation of 'well-qualified and well motivated technical entrepreneurs-innovators', if it was to develop a quality portfolio of 20-plus projects. With Leeds and Wakefield to the immediate north, Bradford and Huddersfield to the north west, and Sheffield, Rotherham and Doncaster to the south and south east, the BBIC was at least within a few hundred thousand of this population criterion. Similarly, within the towns mentioned there were, as of March 1992, three universities and three polytechnics. The cross-territorial interaction might not have worked in practice, however, in quite the way it was envisaged in theory: 8 of the 13 owner/managers interviewed for this study had been living anyway in, or very near, to Barnsley prior to their involvement with the BBIC. Four commuted in from the adjacent west Yorkshire, in three of the four cases from no further than Leeds, 20 miles to the north. Similarly, according to the chief executive of BBIC, the 'spin out' from the nearest university — Sheffield — had not 'worked well'. Although, on his information, there were more than 100 companies working in the laboratories of the University of Sheffield at the time, the world of innovation and entrepreneurship could appear a very risky one from within the confines of a university laboratory, and therefore a world into which university researchers were reluctant to become involved.

Of relevance in relation to the assessment of the impact of the BBIC, was the strongly held view there that the BBIC was 'not really concerned with job creation; we are concerned, rather, with wealth creation'.

The point is important because it goes to the heart of the debate on economic modernisation and regeneration and the nature of post-industrialism. If the desire of economic regeneration projects was to replicate, or nearly replicate, the employment patterns of the past, then the most appropriate methodology of assessment might be to look at the number of new jobs created by a regeneration project, and the cost of those jobs to the taxpayer. There are elements of that type of analysis in this study in some of the individual chapters.

It became increasingly apparent, however, particularly in the 1980s and after, that industrial and economic modernisation — laying the basis for regeneration — might actually bring about a **decline** in employment figures rather than an increase (see, for example, Handy, 1984; Jenkins and Sherman, 1979; Williams, 1986). Machines could replace labour; innovative machinery, using less labour, could replace older machinery, using more labour. This was noted in Chapter Three.

This meant that determining criteria for assessment of the impact of an innovation centre became more problematic than determining the criteria of assessment of some other economic regeneration project. Indeed, where products, processes and services were innovatory, it may be years before their impact on the economy and on society — in terms of jobs 'created' or destroyed; changing work patterns; impact on labour practices or 'industrial' relations; economic modernisation — could be properly assessed. Speculation and weighted human judgement, therefore, entered the process of evaluation of innovation centre efforts.

8.8 Impacts

A central question, however, could be

213

assessed in a quantitatively-specific way: what did the 'innovators-entrepreneurs' think of the operational activities of the innovation centre? Did it enable them to do things they would otherwise have been less able to do? If it did, then the innovation centre can be judged as having had some kind of a positive impact.

Another question that could be addressed more subjectively but nevertheless with some accuracy was: were the products/processes/ services here in any sense 'innovatory'? If they were, then that might provide at least a **chance** of a move towards economic modernisation. If the products/processes/services were judged **not** to be innovatory, however, then the innovation centre could be characterised as being simply another project for the encouragement of small businesses, and the encouragement of people to establish small businesses.

Of less importance, but nevertheless of relevance in judging the **impact** of policy, was the question of which **category** of people were involved in the 'innovation/entrepreneurship'? If it was former coal industry employees involved in the 'innovation/entrepreneurship' then it was possible to make a case for the argument that the impact on the coalfield economy was **direct**: in other words, those formerly involved in traditional industry were now involved in 'new' industry/business. If it was people **other** than those formerly involved, (or who might have become involved, had there been an industry left to become involved in), with the coal industry, then it may be that the impact of the innovation centre was **less** direct. Put another way, if the latter is the case, there is more chance that those involved in innovation would have been **involved in it anyway**, whether or not an innovation centre had existed.

Of the 13 owner/managers of business interviewed, eight believed that being located in the business and innovation centre — as opposed to somewhere else — had been beneficial to their businesses. The overwhelming majority of these — 7 out of the 8 — cited the availability of business advice from the BBIC as being the reason for this. Two cited their view that a 'synergy' had formed between some of the various owner/managers, and that these 'like-minded' people had 'gelled' well together and been able to offer each other help. One cited the knowledge by those at the BBIC of grants and assistance available at European Community level. Overall, then, from the perspective of the individual owner/manager, the BBIC was felt to be a location for this type of business superior to other locations.

Judgement as to whether or not products and services produced by companies within the BBIC were 'genuinely innovative' engendered more difficulty. The largest number amongst the 13 owner/managers interviewed — 5 — were engaging in business which, on their own admission, did not involve any innovation. The second largest number — 4 — believed that the product/service they were engaged in supplying was 'new to the locality': in that sense, in a marginal way, these companies might have been helping to broaden the nature of the economic base in the locality away from its former high dependence on the coal industry. These four listed the services they were offering as: manufacture of analogue optical instruments (measuring equipment); the production of documentary films and advertising; environmental consultancy; an innovatory method for testing for cracks in oil reservoirs, and the provision of training for others who wished to engage in this activity. This latter owner/manager, however, was at the stage of trying to get the business off the ground as of April 1992, and had been at that stage for fourteen months. He was losing money and expected the business to fold within two months if the picture did not change. That left three businesses actually in operation as of April 1992 laying claim to providing a product/service 'new to the locality'. There was some area for disagreement, however, as to whether what these companies were providing was genuinely innovative in the sense outlined earlier: 'invention applied commercially for the first time' (Rowley, 1973, p.33). One of the major projects of the environmental consultancy, for example, was to advise the managers of a storage depot for imported Mercedes Benz vehicles on improving the environment of their business location: this involved advice on building two new ponds, planting something over 4,000 trees, and providing habitats for wildlife (*Barnsley Chronicle*, 3 April 1992). All worthy: but innovative?

Nor did it seem the case that what these environmental consultants were doing was 'new to the locality'. *Yellow Pages* listed one other environmental consultancy firm in the same town; one twelve miles away in northern Sheffield; another fifteen miles away in Huddersfield; and one twenty miles to the north in Leeds (*Yellow Pages*, 1992, p.440). Similarly, the company providing documentary film and film advertisement was engaged in something new to the locality — and in that sense was 'innovatory' within the context of the local economy — but in a wider sense, given that the same services would be available in Manchester 30 miles to the west, the extent of 'innovation' was open to dispute. In any case, unfortunately, that company went into voluntary liquidation in June 1992. Its residual business was taken over by five businessmen who retained the former owner and eight other staff in employment (*Barnsley Chronicle*, 12 June 1992).

Two of the 13 owner/managers interviewed believed that the products/services they were offering were **not** being provided by any other company in the United Kingdom. That would render their products/services 'innovative' at least within the context of the UK economy. Another of the 13 owner/managers reckoned that the product/service his company was supplying was 'fairly unique', which might qualify the product/service offered by that company as being 'innovative' within the context of the UK economy. Of the two, one was offering 'non-contact 3D digiting'. This new product/service would:

'have uses in all industries where quick

accurate measurement of three dimensional shapes is required, for example in the manufacture of moulds used in plastic mouldings and bottle manufacture' (BBIC, undated, b, circa 1991).

The point has to be made, however, that this business was still at the development phase, and employing only one, as of April 1992. It had been at that phase since moving into the BBIC in October 1990.

The other owner/manager who was providing a product/service he believed to be unique within the UK was offering a printing service onto **things** rather than **paper**, employing five.

The owner/manager who put forward the view that his product/service was 'a fairly unique', had a business employing three providing:

'software for the education market ... Software packages include Designer Castles, Designer Villages, Designer Environment and Designer Logic, and they all offer a truly cross-curricular approach to education' (BBIC, undated, b, circa 1991).

There was one other business, however — the winner of the SMART 1 and 2 awards mentioned earlier — which did appear to have a genuine claim to the term 'innovatory' for its product/service. Employing four, including the owner/manager, this company was involved in the development of 'revolutionary techniques for use in mixing powders and solvents for the industrial and pharmaceutical sectors' (BBIC, Autumn 1991), and had been located within the BBIC since 1990. This business had received seed capital from the BBIC itself, and in return the BBIC and company had agreed to a 7 per cent royalty return. This company appeared, as of April 1992, to have considerable potential for expansion and success.

One other company, producing two products for the spark erosion industry (for ultimate use in tool making in the plastics industry), also had a claim to being 'genuinely innovative' —

at least as far as the UK economy was concerned — in relation to one of its products: the 'controller' for spark erosion machinery. No other company in the UK was, to their knowledge, producing this piece of equipment as of early 1992.

This company had started life in the BBIC, and had stayed there three years before moving — though staying within the same town — in early 1992 to find premises more suitable to their requirement for concrete floors. For the owner/managers of this company, being in the BBIC had provided a useful start to business life: in particular, they cited 'respectability' and 'image' for their company as being attributes conveyed by being in the innovation centre. They considered that this had helped them win sales in the early period of their business existence.

In addition, the owner/managers of this company did fit the description 'new entrepreneurs' and were, in the context of the UK economy, innovative. They had previously worked for an established producer of spark erosion generators in the same locality, and had branched off from this employment into the development of a small business producing the 'controllers' for this machine (their previous employers had not manufactured these). Later they began to produce spark erosion generators as well, joining, they thought, about half a dozen other manufacturers in the UK. This company had three full time directors and three other employees as of early 1992. Employment expansion in the future was considered by the owners of this company to be a strong possibility.

The three largest employers within the BBIC, nevertheless, were clearly in the non-innovatory category. The largest employer — employing 25 — had been established since 1983, providing electrical repairs to computer equipment and videos. It had transferred its business from Leeds, 20 miles to the north of the BBIC, in order to obtain 'bigger premises'. Most of its early work had been the servicing of hi-fi equipment for the high street retailers,

218

Boots. It may have been working with modern equipment, but this company did not appear to be 'innovatory': this is similarly indicated by the fact that the move to the innovation centre was prompted by the desire to secure 'bigger premises' rather than to be in a more scientific or technology-oriented environment. A graphic design company within the BBIC which was employing 15 was also similarly open about being non-innovatory. A third company employing 15 in fibre-optic assembly was quite ready to categorise its activities as being non - innovatory: this company, too, had transferred from 20 miles to the north in Leeds.

As for the categorisation of the economic activities of the owner/managers, (or innovator/ entrepreneurs), prior to their involvement in the BBIC, the largest number — 8 of the 13 interviewed — had come from an immediately past background of self-employment, or ownership, or partial ownership, of business. On the definition employed earlier, they had an 'entrepreneurial tradition' and their activities at the BBIC were a continuation of that tradition. Three of the thirteen interviewed had been in managerial positions with other companies prior to their activities in the BBIC: in the steel industry; in the graphic design business; and working for a multinational company abroad.

Only three of the thirteen interviewed had, at any time in the past, worked in the coal industry. Two of these three had been in senior technical positions — one as a geophysicist for British Coal; another as an environmental consultant for the opencast division — and could not on any definition be categorised as meeting the description 'miner'. The remaining owner/manager could: he had been employed as a fitter in the pit, and therefore could be categorised as a former 'miner'. This was the owner of the documentary film and film advertisement company that was mentioned earlier as going into voluntary liquidation in June 1992.

In term of jobs 'created', the impact of the BBIC was modest, as least in relation to the number of jobs **lost** in the coal industry in the nearby vicinity in the mid- and late 1980s and early 1990s. As of March/April 1992, there were about 102 people employed in the small businesses at the BBIC, including the owner/managers of those small businesses, together with 12 staff employed by the BBIC itself. A good number of these, 102 jobs — well over 50 per cent — were 'transferred' jobs from the nearby locality, or form Leeds twenty miles to the north, however. In any case, probably the numbers of jobs 'created' should not be a central part of the analysis here: it was noted earlier that **wealth** creation, rather than simply job creation, was one of the key objectives of the BBIC. If the latter came as a consequence of the former, then that was regarded as being beneficial, but it was the former rather than the latter that was the real aim.

8.9 Conclusions

In principle, the **idea** of a territorially-specific centre aiming at the encouragement of both innovation and entrepreneurship, would be seen as a positive step by most, except perhaps by those who adhere strictly to a non-interventionary free market position in relation to the macro- and micro-economy. The generally favourable disposition of most would be based upon the fact that here was a locality and a local economy that had been involved heavily in the old, traditional industry of coal: it needed to move forward, and the establishment of an innovation centre might at least make some modest contribution to that objective, at a not unreasonable cost to the public purse. At least whatever cost there was to the public purse would be directed towards new industries and businesses, which had a chance for development and long term life rather than, as some would have characterised the coal industry in the 1980s and 1990s, as business/industry where the future looked for the most part gloomy.

The most positive aspect to the BBIC in practice was that the majority of owner/managers

(or 'innovator/entrepreneurs') interviewed —
and these constituted the major owner/managers
in the place in terms of the businesses they
were running — believed that being located in
the BBIC had been beneficial to their business.
The majority identified the reason for this as
being the ready availability of business advice
— on the drawing up of business plans, on
marketing, on helping to put together
applications for grants to other bodies, for
example — within the BBIC; and a lesser, but
still important, identification was that a
'synergy' had developed across and between the
owner/managers in the BBIC.

The findings relating to the usefulness of
business advice can be compared and contrasted
with the case of business advice provision in
Bolsover (see Chapter 4). The immediately
comparable agencies for the provision of
business advice there had been the Rural
Development Commission and British Coal
Enterprise. Bolsover Enterprise Agency, on the
other hand, would provide whatever help and
advice it could to owner/managers, but never set
out, and was not equipped to be, a
professionally and/or technically-oriented
provider of business advice. The contrast
between business advice provision in the
Bolsover Enterprise Park and Barnsley Business
and Innovation Centre was striking. Nine out of
the 28 owner/managers interviewed at Bolsover
had received information/advice from the Rural
Development Commission, but only 2 had found the
RDC's endeavours of any use. Four out of the 28
interviewed had received help or money from
British Coal Enterprise, but only one had found
BCE's endeavours useful.

The research finding that this contrast
pointed to could not be conclusive. Conditions
were different within the two localities
examined, and these differing conditions may
have had a bearing upon the actors (the
owner/managers of small businesses), which
reduced any potential for a direct comparison.
Bolsover was a small town in north Derbyshire,
for example, whereas Barnsley was a fairly big
town in south Yorkshire; the physical

environment provided by the BBIC in Barnsley was modern, clean and to a sophisticated specification, whereas the physical environment in Bolsover was comprised of recycled British Coal buildings designed originally for a different business use, and within yards of a still operating colliery and coking plant.

Nevertheless, within the context outlined above, what could be argued was that the fostering of small business ownership and small business promotion in Bolsover represented more of an 'entrepreneurship in vacuity' situation than did the equivalent operations in Barnsley. The major form of business advice that was offered in Bolsover was by nationally-based organisations — British Coal Enterprise and the Rural Development Commission — who could devote only part of their attention to that locality. In the Barnsley Business and Innovation Centre, by contrast, the staff had one set of territorially-captive (voluntarily so) owner/managers or 'entrepreneur/innovators' to concentrate on only. It would appear therefore that the provision of on-the-spot territorially-specific business advice would have more chance of meeting a favourable response from owners/managers of small businesses (upon which the regeneration strategies studied in both Barnsley and Bolsover depended), than on a strategy which was closer to 'entrepreneurship in vacuity'.

The record of the BBIC on 'genuine innovation', upon which rested the prospects of the BBIC contributing to a technological modernisation of the local economy, was mixed and probably less positive than its record on the provision of business advice. Of the four owner/managers out of the 13 owner/managers interviewed that did lay claim to being involved in some activity genuinely innovative within the context of the UK economy, two of these were still at the development phase as of April 1992. Over the third ('printing on to things rather than paper'), there appeared to be a question mark over the extent to which it was genuinely innovative. In four years of its existence then, between 1980 and 1992, there was only one company up and running which could lay claim to being a genuinely innovative firm: it has to be said that progress to a high technology local economy via this route would take a long time. Moreover, in relation what is often at least a long term objective of regeneration efforts — employment creation on a significant scale — it could hardly be said that the prospects for that looked good from the perspective of April 1992.

The real point that needed addressing, because of its bearing upon the operational effectiveness of a body such as the BBIC, related to the nature of the socio-economic environment within which it was operating. For an innovation centre to be operationally successful, it would need to be within a socio-economic environment within which innovation and entrepreneurship, if not exactly the norm, at least had a presence and an existence. Where it did not, an innovation centre would be working against the grain of a 'local economic culture'. Evidence suggested that this factor was apparent in Barnsley: 2 (companies 1 and 2 in Tables 8 and 9) out of the 4 owner/managers with businesses which either were in actuality, or at least which their owner/managers believed were, genuinely innovative, commuted to the BBIC from west Yorkshire. A third (company 4 in Tables 8 and 9), had moved to south Yorkshire from west Yorkshire to get governmental assistance with his project. A fourth (company 4 in Tables 8 and 9), though having run a business in Barnsley immediately prior to his activities with BBIC, had originated in Plymouth. The chief executive of the BBIC offered the view that for the BBIC, 'finding the right people was the most difficult thing', and that the 'local culture was not conducive to innovation'.

If that was the case, the debate then moved on to the much longer-term and difficult question: how was an 'innovatory culture' to be developed? Mining communities are well known for developing their own distinctive cultures over generations, centred around the workplace, the welfare club, the family, the union, sometimes the sports and recreation facilities. Such cultures began to be jolted exogenously in the 1960s, 1970s and 1980s — by pit closures and the consequent dispersal of the workforce, or by the diminution of the importance of the union, or the growth of employment in alternative industries — but, because of their internal strength, such cultures could not and would not disappear overnight. As faced those who sought to develop an 'enterprise' culture as a regeneration strategy in localities traditionally associated with coal mining, (see earlier chapters), the development of an 'innovative' culture may not be an easy task. The obvious starting point would be by a greater input of technology into the school curriculum (which did begin to happen under the Conservative governments of the late 1980s and early 1990s), but the prospects of this having a major impact on the culture, attributes, and capabilities of the majority of people in former coalfield communities seemed remote from the perspective of the early 1990s.

To summarise then, the **idea** of an innovation centre in a locality of coal mining decline would receive a positive endorsement from most politicians and economists, even though the majority of funding in this case came from the public purse. The impact of this particular centre appeared to have been less than 'direct', on the definition applied above, in the sense that most of the 'clients' (owner/managers) were self-employed in their immediately prior working existence to the BBIC, rather than being people who had worked in the coal industry, or who might have expected to. That did not necessarily, however, present a problem: industrial pioneers need not, and have not usually, come from within the ranks of the workers themselves. Nevertheless, there appeared to be little net addition here to the 'stock' of those engaging in 'entrepreneurial' activity. A problem facing this particular innovation centre was that 'innovation' had not had a prominent role in the predominant 'local economic culture'. It would appear that relying on a body such as an innovation centre in a locality formerly associated with coal mining would result in only a very modest contribution

towards the technological modernisation of a
local economy, and only then over the very long
term. To say that it was modest and long term,
however, did not mean that it was not worth
doing: for the prospect of technological
modernisation, even if only in the long term,
had at least the chance of seeing the
development of a sustainable local economy built
upon the new rather than the old. So at the
very minimum, this innovation centre was trying
to be something more than a place simply within
which ordinary small businesses were fostered
and encouraged.

	Business	Job Level	Genuine Innovation	Centre Useful?	If so, why?
1	Non-contact 3D Digiting	1	Yes. Nobody else in UK doing it	Yes	Synergy
2	Mixing chemicals	4	Yes. Winner of SMART award	Yes	Marketing help Knowledge of EC
3	Computer software for education	3	Yes. Product 'fairly unique'	Yes	Technical/business advice Knowledge of EC funding
4	Printing onto things, not paper	5	Company believes so	Yes	Technical help Synergy
5	Oil reservoir testing & technology training	1	New to locality	Not really	n/a
6	Documentary Films/Adverts	14	New to locality	Yes	Business advice
7	Optical Instrument Measuring Equipment	Low level No data	New to locality	No	n/a
8-11	Company name registered. Same owner as above. No additional employment				
12	Environmental Consultancy	3	Claimed to be new to locality	Yes	Business advice provision
13	Electrical Repairs	25	No	Yes	Business advice provision
14	Marketing Consultants	3	No	No	n/a
15	Graphic Design	15	No	No	n/a
16	Fibre Optic Assembly	15	No	Yes	Business advice provision
17	Electronic products for mining	7	No	No	n/a
18	Laboratory tests logged on computer	1	Unlikely	Not known	n/a
19	Computer Firm	1	Not known	Not known	n/a
20	Not known	2	Not known	Not known	n/a

	Company in Innovation Centre Since?	Former Employment of Owner/Manager	Locational Origin/Residence of Owner/Manager	Other Comments
1	October 1990	Self-employed from home	Commuted from West Yorkshire	Business at development stage
2	1990	Owned a similar company in Leeds	Commuted from West Yorkshire	Business at development stage. BIC had 7% royalty stake
3	March 1990	Self-employed from home	Moved 40 miles to Barnsley to obtain DTI help	Received regional investment grant from DTI
4	Late 1991	Small business ownership	South of England but last business in South Yorks	
5	Feb 1991	Geophysicist with British Coal	Commuted from West Yorkshire	Development stage
6	1988	Fitter in coal mine	Local	Liquidation July 1992
7	Late 1989	Part owner of small business	Local	
8-11	Company name registered. Same owner as above. No additional employment			
12	September 91	Environment consultant to British Coal	Local	
13	August 1989	Transferred same business from 20 miles to north	West Yorkshire	
14	March 1992	Steel industry	Local	'Would have working in area anyway'
15	March 1992	Another Graphic Design Co	Local	
16	April 1992	Transferred same business from 20 miles north	West Yorkshire	
17	January 1992	Transferred same business from within 5 miles	Local	
18	Not known	Not known	Not known	
19	Not known	Not known	Not known	
20	Not known	Not known	Not known	

227

9   Conclusions

The major conclusions to this work are contained within the individual chapters. All that is intended here is a summary and overview.

The first concluding point should be a recognition of the extent of the problem of, and the problems caused by, deindustrialisation in the coalfield communities. The rapidity of economic change has been devastating for communities which have relied for generations for their economic livelihood on one major industry.

Secondly, factors very often common to these communities — infrastructural isolation; lack of transferable skills; often low educational performance — militated to hold back **against** 'spontaneous economic recovery': in other words, economic recovery without outside (governmental/quasi governmental) assistance. Herein was located the economic justification for government, or quasi-government, led regeneration efforts.

Thirdly, until the dramatic events of

October 1992, the coalfield communities achieved only **partial** success during the 1980s and early 1990s in placing on the central government's 'policy agenda' their 'needs/demand/wants' brought forth by deindustrialisation: the 'need' for economic regeneration efforts, for example. For many years, there was no attempt even by the NUM itself to place on the political agenda the issue of the problems caused by pit closures, even though evidence suggests that from at least the late 1960s onwards pit closures were promoting serious economic and social problems in specific coalfield localities. This lack of effort by the NUM to place the issue of the problems caused by pit closures was largely a consequence of, firstly, a widespread acceptance within the union of a dichotomy between the 'industrial' and the 'political'. Only the former — involving negotiations over wages, working conditions, and so on — was seen as a legitimate activity for the union to pursue. The latter — which included decisions on the size of the industry and therefore pit closures — was a task for politicians. Hence, rather than fight against pit closures, the NUM invested its faith in a political solution to pit closures which they believed would follow victory in a general election by the Labour Party. That strategy, of course, failed in the 1960s as Harold Wilson's Labour government pursued a draconian pit closure programme. By the time of the 1974-79 Labour government the salience of the issue of pit closures had waned considerably, given the changed energy environment. The second factor militating against the issue of the problems caused by pit closures figuring on the political agenda was the existence for much of the post-Second World War era — at least until the end of the 1960s, and in some places into the mid-1970s — of full employment. Displaced coal miners could find paid employment in other industrial sectors, and in many cases were happy to get out of the pit.

A fourth conclusion is that, for both pragmatic and ideological reasons, central government during the 1980s and early 1990s favoured carefully targeted, territorially-specific regeneration efforts — British Coal Enterprise, enterprise agencies, enterprise zones, task forces, City Challenge schemes — over a less territorially-specific 'blanket' regional policy.

As regards policy-making mechanisms, if there were policy communities (as these were defined earlier) in relation to economic regeneration in existence, at least prior to October 1992, the strongest that can be said is that these were fragmented and transient. They did not exhibit the permanence nor the internal coherence attributed to them by some of the adherents of the 'policy community theory' such as Heclo and Wildavsky (1981).

A central theme in this work was that the **choice** of particular regeneration activities — **as against other potential regeneration activities** — was not a reflection of political neutrality, born out of the workings of objective economic analysis. Rather, the particular regeneration schemes chosen reflected the **ideological** aims of central government during the time. In some cases, the ideological objective was to by-pass local authority channels of potential activity. Local authorities might, during an earlier time period, have been allocated a more central role in policy efforts. The enterprise zone and the task force examined here represented specific examples of the by-passing of local authorities in economic regeneration policy. A second objective pursued in these regeneration schemes for, it can be argued, as much ideological as strictly economic reasons, was the aim of bolstering and adding to the ranks of the small business sector. Where local authorities pursued the same kind of strategy, a case can be made for saying that this might not otherwise have been their preferred strategy, but money resources were tight and getting tighter throughout the 1980s and early 1990s: this in itself might have precluded other, more comprehensive, regeneration activities by local authorities.

The emphasis on bolstering the small business sector as a potential engine for economic regeneration chimed neatly with central government's professed desire during this period to foster an 'enterprise culture' within society and the economy. If it could be demonstrated that people were **voluntarily** joining this sector — becoming 'new entrepreneurs' — that might have been one sign of an 'enterprise culture' successfully emerging. If people were joining the sector for other reasons — such as high unemployment, and therefore a lack of opportunities with existing employers — then that would represent a form of coercion as a factor working to expand the small business and self-employment sector. In such circumstances, the idea that a robust 'enterprise culture' was emerging could hardly be supported. The evidence on whether or not a genuine 'enterprise culture' was emerging, from Carcroft and Bolsover for example, was mixed: there **was** some 'new entrepreneurship' but a high percentage of people had been to some extent coerced into it in an effort at economic survival. In any case, it is not clear that interviewees would in all cases voluntarily offer an explanation for self-employment the reason that they could not get a job elsewhere, even if that were the case. It should be noted that a high percentage of those taking advantage of small-business promotion facilities had, in any case, a career history of being involved in the ownership or running of small business: their activities did not therefore denote a net addition to the numbers involved in the small business sector. **Very** few former coal miners had become involved in the small business ownership sector: for them, the stimulation of this sector, even where successful, was not a solution to their individual unemployment problem. A limited number of former coal industry employees did find success or partial success in this direction, though more often than not these came from the supervisory, technical or management sector of the industry. What was also evident was that in many coalfield localities, the 'local economic culture' militated **against** the stimulation of the small business sector and 'new entrepreneurship'. The consequence of that was that people coming newly to small business ownership, self-employment or 'new entrepreneurship' welcomed and benefited from the provision of business advice. The higher the technical and professional standard of this, the better. Evidence from Bolsover (Chapter 4) and Barnsley (Chapter 8) corroborated this, where a comparison between a moderate, (Bolsover), and a high (Barnsley) standard of provision could be made.

This study examined regeneration efforts which had objectives which could be divided into two: employment creation; and 'modernisation' of the local economy. Some projects combined both objectives (the Doncaster Task Force, for example); others were primarily focused on one: the South Kirkby enterprise zone on employment creation, for instance, or the Barnsley Business and Innovation Centre on 'modernisation' of the local economy. In the 'modernisation' category could be placed any efforts at moving a local economy from a phase of being characterised by 'traditional' industry, through to a phase of higher technology or producer service-orientation. This would include training efforts which would potentially, if successful, provide **individuals** at least with a route from one economic phase to another.

Clearly, from the perspective of the early 1990s, the coalfield and former coalfield localities needed **both** the creation of paid employment **and** economic modernisation. In a reflection of the only partial acceptance on to the 'policy agenda' of the issue of the problems caused by pit closures, the policy response from central government, and more importantly, the policy outcome, in terms of economic regeneration success, was only partial and modest. It was possible, of course, that the economic problems of former coalfields were so severe that even if the issue of the problems caused by pit closures **had** been fully accepted on to the government's 'policy agenda', success in economic regeneration efforts would have been elusive. Although there were some training initiatives aimed specifically at people in the coalfields — the Doncaster Task Force initiative was one, and Barnsley and Doncaster Training and Enterprise Council made other special efforts in the Dearne Valley in south

Yorkshire (*Barnsley Independent*, 21 July 1992)
— it could be argued that, at the time of
writing, there was a strong case for further and
enhanced efforts in this direction. This may
have served to help people where there was
evidence that young, and sometimes not so young,
people had **never** worked. That some young people
had never worked was clearly the case in the
Dearne Valley in south Yorkshire, for instance
(Coopers and Lybrand Deloitte and Sheffield City
Polytechnic, 1990).

Any temptation to brand any of the case
studies examined as 'policy failures', without
other qualification, should be resisted. If
they had their shortcomings, they all achieved
some benefit: buildings were converted;
derelict land reclaimed; a limited number of
people gained retraining; for example.
Moreover, what could not be known, was what
would have happened in the **absence** of the
regeneration schemes examined. It may well have
been that the economic situation in the relevant
localities would have been much worse. For
example, it may have been the case that without
the industrial subsidies detailed, the firms on
South Kirkby enterprise zone would have
performed less well financially and would not,
as a consequence, have been able to provide the
employment levels that they did provide.
Another finding from this study was that the
combined impact of deindustrialisation in coal,
and economic regeneration efforts, was modifying
patterns of labour organisation. In other
words, trade unionism was weakening and
declining. Partly this was simply a consequence
of the decline of the 100 per cent unionised
coal industry, within which the NUM developed,
at times, industrial power and the willingness
to use that power. Partly it was because the
small businesses that were being encouraged as
part of the regeneration efforts were not
appropriate economic vehicles to be unionised:
there would not be much point in the self-
employed striking against themselves. Partly it
was because where companies were inwardly
investing they were often signing single-union
agreements with non-militant trade unions.

In 1991, for example, subsidiaries of two

companies inwardly investing into the Barnsley
area — Koyo Bearings (Europe) Ltd (Japanese
owned multinational) and Kostal UK (German owned
multinational) — signed single union agreements
with the Electrical Electronic Telecommunication
and Plumbing Union ((EETPU) which earlier had
been expelled from the TUC (*Barnsley Chronicle*,
13 December 1992).

In a wider study of regeneration efforts in
the coalfield and former coalfield areas,
attention would have been paid to the
effectiveness of the promotion, and the impact,
of inward investment efforts by local
authorities and other bodies in these
localities. Coalfield local authorities pursued
this policy aggressively in the 1980s and 1990s.
This could be judged from the examples above and
others mentioned within the text. That this was
not prioritised for this study reflected the
fact that attempts to attract inward investors
was not particularly distinctive to coalfields:
virtually everybody was at it throughout the
1980s and 1990s. Nevertheless, there remained,
as of 1992, further research to be carried out
on the impact — in employment terms, in terms
of industrial structure and labour organisation
— of firms inwardly investing into the
coalfield and former coalfield localities.

A second area where, as of 1992, research
remained to be carried out in depth was on the
impact of the conversion of former colliery
sites into industrial estates. Sometimes this
connected in with the above as former colliery
sites became locations for inwardly investing
companies: the Japanese Koyo company mentioned
above, for example, came to the site of the
former Dodworth colliery near Barnsley in 1992.
Other former colliery sites were turned into
'business parks', and became home to more
indigenous firms, such as the former Redbrook
colliery near Barnsley, for example; or became
home to other economic activities. The site of
former North Gawber colliery near Barnsley,
became home to an 8,000 square foot supermarket
in 1992, for instance (*Barnsley Chronicle*, 17
July 1992). Other examples of colliery sites
being turned into industrial estates, or turned
over to other economic or social uses abounded:

the site of the former Lindby colliery in
Nottinghamshire was being converted into
industrial, housing and public open space as of
1992; the site of the former Yorkshire Main
colliery site near Doncaster was accommodating a
supermarket, petrol station and private housing
as of that date.

The economic and social problems caused by
pit closures in the 1980s and early 1990s were
deep and probably long-term. To stand any
chance of success, regeneration efforts would
have to be extensive and intensive, and take a
variety of forms, if different people in the
coalfields were to benefit from them.
Encouraging entrepreneurship, for example, had
the advantage of being inexpensive and easy to
organise, but not all, (or even many), in
coalfield areas would see themselves becoming
'entrepreneurs'. Others might need paid
employment provision and/or training if they
were to benefit from economic regeneration
efforts in the coalfields. Certainly, from the
perspective of 1992, economic regeneration
efforts in the coalfields, if they were to have
a prospect of serious success in reversing the
adverse economic consequences of
deindustrialisation, would require a long-term
commitment of both financial and political
capital from central government. It appeared in
the early 1990s, however, as if central
government had attached a greater policy
priority to contraction in the coal industry to
bring it to self-standing financial viability,
and to make what remained of it saleable to the
private sector. That position was not even
threatened, until October 1992, and even after
that date it seemed as if the Conservative
central government's priorities were unchanged.

Appendix 1

Postal Questionnaire to Companies on Langthwaite Grange Industrial Estate, South Kirkby Carried out September to December 1990

1   Name of Company:

2   What is the nature of your business?

3   When did your company establish operations in the Enterprise Zone (month and year).

4   How many people does your company employ presently?

5   How many people did your company employ at the commencement of its operations in the Enterprise Zone?

6   Do you consider your move onto the Enterprise Zone a relocation of an existing business?   Yes/No

7   If it is a relocation of an existing business, where was your company located previously?

8   Did your company create any extra jobs as a result of the relocation?   Yes/No

9   If yes, how many?

10   Did your company shed any jobs on relocating to the Enterprise Zone? Yes/No

11   If so, how many?

12   Would your company have located in this geographical area had an Enterprise Zone site not been available?   Yes/No

13　Would your company have been able to employ as many workers as it does employ if an Enterprise Zone site had not been available?　　　　　　　　　　　　　Yes/No

14　What percentage of the company's Enterprise Zone workforce is male?

15　What percentage of the company's Enterprise Zone workforce is female?

16　What percentage of the company's Enterprise Zone workforce works full-time?

17　What percentage of the company's Enterprise Zone works part-time?

18　Has your product range expanded or changed significantly since moving onto the Enterprise Zone?　　　　　　　　　　　Yes/No

19　If so, in what ways?

20　Has your company introduced new machinery or management techniques since moving onto the Enterprise Zone?　　　　　　Yes/No

21　If yes, can you briefly describe these?

22　Have the company's operations changed significantly in some other way since moving onto the Enterprise Zone?　　Yes/No

23　If yes, in what way?

24　If your Enterprise Zone business is a relocation: has the profitability of your operations improved since moving onto the Enterprise Zone?　　　　　　　　　Yes/No

25　If yes, by how much?

26　If your Enterprise Zone business is an **expansion** of an existing business, is your Enterprise Zone operation more profitable than your other business operations?Yes/No

27 Which companies do you consider to be your major competitors?

28 Have the Enterprise Zone advantages, such as absence of business rates, helped your company become more competitive?   Yes/No

29 Is a trade union, or trade unions, recognised by your company?   Yes/No

30 If yes, which?

31 Has your company received exemption from Development Land Tax?   Yes/No

32 If yes, to what extent, in money terms?

33 Has your company benefited from capital allowances in relation to corporation or income tax for capital expenditure on the construction, extension or improvement of industrial and/or commercial buildings?
                                          Yes/No

34 If yes, to what extent, in money terms?

35 What is the average weekly wage of your employees, before tax and other stoppages?

36 What were your main reasons for moving onto the Enterprise Zone?   Have your expectations been fulfilled?

Appendix 2

Postal Questionnaire to Selected Companies in Barnsley on employment structure and characteristics
Carried out February 1992

March 1992

Dear Sir/Madam

I am carrying out some research into the nature and structure of the economies of coalfield and former coalfield areas. I am sending this letter to selected companies in the Barnsley locality to seek information in a number of specified areas. I wonder if you would be so kind as to provide information on the following:

1   What is the nature of your business?

2   When did your business operations in Barnsley begin?

3   How many people work in your organisation?

4   In relation to question 3 above, can these people be sub-divided numerically into:

    a   managerial;
    b   white collar/office staff;
    c   other technical staff such as foremen, supervisory on the shop floor;
    d   blue collar/manual.

5   What percentage of your total workforce is:

    a   male;
    b   female.

6   What percentage of your **managerial** workforce (where this category exists) is:

    a   male;
    b   female.

239

7   What percentage of your **white collar/office staff** (where this category exists) is:

a   male;
b   female.

8   What percentage of your **other technical staff** (where this category exists, as defined above) is:

a   male;
b   female.

9   What percentage of your **blue collar/manual workforce** (where this category exists) is:

a   male;
b   female.

10   Is a record kept of the immediate former occupations of your employees?

11   Is a record of the total number of previous jobs (or career history) of your employees?

12   If information relating to questions 10 or 11 is available, how many of your employees formerly worked in the coal industry?

13   If information relating to the above question is available, can it be subdivided into:

a   how many of your employees formerly employed in the coal mining industry are employed by yourselves in managerial positions?

b   how many of your employees formerly employed in the coal mining industry are employed by yourselves in white collar/office staff positions?

c how many of your employees formerly employed in the coal mining industry are employed by yourselves as 'other technical staff' as defined in question 4?

d how many of your employees formerly in the coal mining industry are employed by yourselves in blue collar/manual positions?

14 If there are any of your employees which fit the category listed at 13(a), how many of them were:

a technical, managerial or supervisory grades in the coal industry (include here deputies, overmen, supervisory grades at administrative/managerial level; surveyors, tracers and similar);

b skilled underground workers such as electricians, welders, fitters;

c faceworkers; other underground workers; manual surface workers; drivers; colliery haulage workers and similar;

d junior clerical workers.

15 In which town/city/country is the headquarters of your company located?

Appendix 3

Local Authority Members of Coalfield Communities Campaign as of April 1991

Authority   **Political Control as of 1991**

Scotland

1	Central	Labour
2	Clackmannon	Labour
3	Clydesdale	Labour
4	Cumnock and Doon	Labour
5	Dumfries and Galloway	Labour
6	Dunfermline	Labour
7	East Lothian	Labour
8	Edinburgh	Labour
9	Fife	Labour
10	Kirkcaldy	Labour
11	Kyle and Carrick	Labour
12	Lothian	Labour
13	Midlothian	Labour
14	Monklands	Labour
15	Motherwell	Labour
16	Nithsdale	Labour
17	Stirling	Labour
18	Strathclyde	Labour
19	West Lothian	Labour

North East

20	Alnwick	Labour
21	Blyth Valley	Labour
22	Castle Morpeth	No overall control (Independent, Labour and Liberal Democrat equal representation)
23	Chester-le-Street	Labour
24	Derwentwide	Labour
25	Durham City	Labour
26	Durham County	Labour
27	Easington	Labour
28	Newcastle upon Tyne	Labour
29	North Tyneside	Labour
30	Northumberland	Labour

31	South Tyneside	Labour
32	Sunderland	Labour
33	Teesdale	Independent
34	Wansbeck	Labour

North West

35	Knowsley	Labour
36	Salford	Labour
37	St Helens	Labour
38	Warrington	Labour
39	Wigan	Labour

Yorkshire

40	Barnsley	Labour
41	Doncaster	Labour
42	Kirklees	Labour
43	Leeds	Labour
44	Rotherham	Labour
45	Sheffield	Labour
46	Wakefield	Labour

Midlands

47	Ashfield	Labour
48	Bassetlaw	Labour
49	Bolsover	Labour
50	Cannock Chase	Labour
51	Chesterfield	Labour
52	Derbyshire	Labour
53	Gedling	Conservative
54	Hinckley and Bosworth	Conservative
55	Leicestershire	No overall control. Tories biggest party
56	Mansfield	Labour
57	Newark and Sherwood	Labour
58	Newcastle-under-Lyme	Labour
59	North East Derbyshire	Labour
60	North Warwickshire	Labour
61	North West Leicestershire	Labour
62	Nottingham	Labour
63	Nottinghamshire	Labour
64	Nuneaton and Bedworth	Labour
65	South Derbyshire	Labour
66	Staffordshire	Labour

243

67	Stoke-on-Trent	Labour

Wales

68	Blaenau Gwent	Labour
69	Brecknock	Independent
70	Carmarthen	Independent
71	Clwyd	Labour
72	Cynon Valley	Labour
73	Dinefwr	Labour
74	Dyfed	No overall control - Independents biggest
75	Gwent	Labour
76	Islwyn	Labour
77	Llanelli	Labour
78	Lliw Valley	Labour
79	Merthyr Tydfil	Labour
80	Mid Glamorgan	Labour
81	Neath	Labour
82	Ogwr	Labour
83	Port Talbot	Labour
84	Rhondda	Labour
85	Rhuddlan	No overall control - Labour biggest party
86	Thymney Valley	Labour
87	Swansea	Labour
88	Taff-Ely	No overall control - Labour biggest party
89	Torfaen	Labour
90	West Glamorgan	Labour
91	Wrexham maelor	Labour

Sources: Coalfield Communities Campaign (undated, circa April 1991); Municipal Yearbook (1992).

Business Rates Applicable in City of Wakefield
Metropolitan District Council Administrative
Area, 19981/2 to 1990/91

Financial year	Business Rate Applicable in City of Wakefield Metropolitan District Council Administrative Area
1981/82	£1.341
1982/83	£1.806
1983/84	£1.848
1984/85	£1.991
1985/86	£2.251
1986/87	£2.825
1987/88	£3.0672
1988/89	£3.3652
1989/90	£3.6621
1990/91	£0.348 (National Uniform Business rate)

Analysis of Subsidies Paid to South Kirby
Enterprise Zone Companies

(As of October 1990)				
Co/Nature of Business	Moved onto EZ when? (i)	Rateable Value (ii)	Workers (iii)	Estimated Subsidy £
1 Clothing Manufacturer	Aug 90 (lr)	2,242 (e)	370	195
2 Frozen Food Distributor A	Already there	102,805 (v)	150	2,406,074
3 Frozen Food Distributor B	Already there	2,263 (v)	250	54,734
4 Clothing Manufacturer A	Start of EZ designation	7,000 (v)	193	168,228
5 Sports Shoe	Already there	10,125 (v)	158	243,501
6 Packer	Start of EZ designation	21,638 (v)	113	376,092
7 Transport Company A	Oct 84	14,222 (v)	100	258,784
8 Mining Engineers	One site already there; another Dec 88	9,659 (v) 7,500 (v) respectively	87 222	280,404
9 Chemicals A	Nov 87 (lr)	4,201 (v)	70	43,419
10 Food Producer	Start of EZ designation	13,100 (v)	66	310,987
11 Meat Packer	Jan 84	2,500 (v)	63	47,069
12 Office Furniture	One site Oct 87 another Aug 89	11,447 (e) 5,723.50 (e) resp	60	146,826
13 Printer A	Oct 82	2,242 (v)	50	153,548
14 Dyer	One site Oct 82 another Apr 85 another Jan 88 (lr)	547 (v) 5,638 (v) 5,638 (v)	38	164,367
15 Trainers	May 85	742 (v)	36	12,624

Co/Nature of Business	Moved onto EZ when? (i)	Rateable Value (ii)	Workers (iii)	Estimated Subsidy £
16 Glass Recyclers A	Oct 7 (lr)	8,600 (v)	35	91,082
17 Condom Distributor	Start of EZ designation	4,500 (v)	35	96,698
18 Food Packer	Start of EZ designation (lr)	15,263 (v)	32	327,922
19 Engineers A	Already there	3,800 (v)	19	90,210
20 Engineers B	Apr 89 (lr)	701 (e)	19	3,990
21 Security Firm	May 86 (lr)	547 (e)	15	7,920
22 Coils Co	Jan 89	3,800 (e)	15	24,827
23 Freight Co Ltd	Sep 89	5,000 (e)	15	20,831
24 Pet Food	Oct 86	1,248 (v)	10	16,600
25 Joinery	Jun 86 (lr)	1,930 (v)	9	27,489
26 Sweet Co	Apr 87	547 (v)	9	6,631
27 Mushroom Equipment	Already there	2,525 (v)	9	59,942
28 Tea Distributor	Sep 86 (lr)	8,722 (v)	8	120,123
29 Coal Importers	Apr 85	2,201 (v)	7	37,858
30 Engineers C	Already there	363 (v)	6	8,617
31 Animal Co	Sep 82	1,641 (e)	6	36,255
32 Glass Recyclers B	Already there	805 (v)	5	19,109
33 Storage Co	Jan 89	5,000 (e)	5	32,667
34 Lampshade Co	Start of EZ designation	1,641 (e)	5	38,773
35 Drilling Co	Jul 88	194 (e)	5	1,540
36 Car Dismantler	Dec 87	230 (e)	4	2,318
37 Sign Maker	Oct 86	547 (v) and 351 (v)	4	12,156
38 Injection Moulding	Sep 89	266 (e)	4	2,409
39 Steam Cleaners	Already there	223 (v)	3	5,394

Co/Nature of Business	Moved onto EZ when? (i)	Rateable Value (ii)	Workers (iii)	Estimated Subsidy £
40 Panel Beaters	Oct 83 (lr)	347 (v)	3	6,980
41 Engineers D	Oct 85	266 (e)	3	4,276
42 Chemicals B	Apr 87	701 (v)	3	8,499
43 Resin Co	Oct 88	2,500 (e)	3	18,437
44 Mining Engineers	Sep 89 (lr)	1,094 (e)	3	4,892
45 Sunbed Manufacturers	Oct 88 (lr)	1,930 (e)	3	14,233
46 Carpet Co	Start of EZ designation	2,097 (v)	2	49,782
47 Hire Co	Nov 89	266 (e)	2	1,514
48 Motor Spares	Feb 83	547 (v)	2	11,591
49 Motor Cycle Co	Oct 87	266 (v)	2	3,225
50 Coal Merchant	Apr 88 (lr)	430 (e)	1	3,895
51 Printer B	Oct 89	266 (e)	1	1,027
52 Welding Co	Oct 87	266 (e)	1	2,817
53 Equestrian Supplies	Nov 87	266 (e)	1	2,749
54 Engineers E	Mar 87	266 (v)	1	3,289
55 Haulage Co	Information not available	266 (e)	1	nominal 500
a Double Glazing	Start of EZ designation; Closed Feb 90	23,500 (v)	0	503,049
b Clothing Manufacturer C	Start of EZ designation; Closed Apr 90	13,700 (v)	0	301,393
56 Paper Co	Feb 83; closed by Oct 90 but site still in business use	2,472 (v)	0	53,870
c Window Co	Jul 82; closed Dec 89	1,480 (v)	0	28,109

Co/Nature of Business	Moved onto EZ when? (i)	Rateable Value (ii)	Workers (iii)	Estimated Subsidy £
d Chemicals C	Start of EZ designation; closed Apr 89	3,800 (e)	0	69,740
e Misc 1	Operated short period	2,084 (v)	0	nominal 2,084
f Misc 2	Operated short period	500 (v)	0	nominal 500
g Engineers F	Operated short period	266 (v)	0	nominal 500
h Misc 3	Operated short period	266 (e)	0	nominal 500
i Sweet Co 2	Operated short period	547 (e)	0	nominal 2,000
j Construction Co	Operated short period	500 (e)	0	nominal 500
k Book Co	Operated short period	100 (e)	0	nominal 100
57 Building Co	Oct 83 onwards (a store)	266 (v)	0	nominal 5,351
l Misc 4	Operated short period	500 (e)	0	nominal 500
m Diesel Co	Operated short period	500 (e)	0	nominal 500
n Insulation Co	Operated short period	500 (e)	0	nominal 500
o Concrete Co	Operated short period	500 (e)	0	nominal 500
p Stationery Co	Operated short period	500 (e)	0	nominal 2,000
q Misc 5	Operated short period	500 (e)	0	nominal 2,000
r Misc 6	Operated short period	500 (e)	0	nominal 500
Total			2,120	6,870,115

249

Notes to Appendix 5

i)   'lr' represents 'local relocation':   usually from within 5 miles of the enterprise zone site; 'already there' means company was present on the industrial estate prior to its designation as an enterprise zone; 'start of EZ designation' means company moved onto industrial estate at the beginning of its designation as an enterprise zone;

ii)   'v' represents a rateable value for property verified with the City of Wakefield Metropolitan District Council; 'e' represents an estimate of rateable value for property based upon a comparison of the business property in question with a similar size and type of business property where a verified rateable value **did** exist;

iii)   employees in full-time paid employment or the equivalent;

iv)   companies still trading as of October 1990 are designated by number;   those which had ceased to trade by then by letters.

References

Newspapers

Barnsley Chronicle (1991)   26 July
Barnsley Chronicle (1991)   2 August
Barnsley Chronicle (1991)   20 September
Barnsley Chronicle (1991)   4 October
Barnsley Chronicle (1991)   8 November
Barnsley Chronicle (1991)   13 December
Barnsley Chronicle (1992)   'Whitbread cash
                                     boost for BBIC',
                                     14 February
Barnsley Chronicle (1992)   'Councillor
                                     attacks   £1.5m
                                     partnership cash
                                     commitment',
                                     13 March
Barnsley Chronicle (1992)   'Bid to improve
                                     landscape'
                                     3 April
Barnsley Chronicle (1992)   'Consortium takes
                                     over   failed
                                     production

Barnsley Chronicle (1992)   company', 12 June 17 July

Barnsley Chronicle (1993)   'Over 170 jobs to go as BC's Shafton Workshops close,' 11 June

Barnsley Independent (1992)   21 July

Daily Telegraph (1979)   14 June

Daily Telegraph (1991)   'Coal mining areas are "missing out",' 26 July

Doncaster Star (1987)   28 April

Doncaster Star (1987)   29 April

Doncaster Star (1990)   15 August

Doncaster Star (1990)   7 September

Doncaster Star (1991)   23 November

Financial Times (1985)   8 January

Financial Times (1985)   13 May

Financial Times (1986)   27 September

Financial Times (1987)   'NUM leaders told 1,300 jobs to go in pit closures', 17 October

Financial Times (1988)   4 February

The Guardian (1984)   15 November

The Guardian (1984)   29 November

The Guardian (1985)   15 May

The Guardian (1985)   6 June

The Guardian (1986)   16 May

The Guardian (1991)   5 August

The Guardian (1991)   2 September

The Guardian (1991)   5 September

The Guardian (1991)   '900 pit jobs to go', 12 September

The Guardian (1991)   11 October

The Guardian (1991)   'Coal Chairman fights carve up of pits industry', 17 October

The Guardian (1991)   '300 jobs cut with more under threat', 29 November

The Guardian (1991)   'Heseltine memo accuses ministers

The Guardian (1992)   over EC aid', 18
                                             December
The Guardian (1992)   9 April
The Guardian (1992)   29 July
The Guardian (1992)   18 September
The Guardian (1992)   26 September
The Guardian (1992)   14 October
Hemsworth and South
Elmsall Express (1982)   29 July
Hemsworth and South
Elmsall Express (1986   5 June
Hemsworth and South
Elmsall Express (1987)   27 May
Hemsworth and South
Elmsall Express (1988)   16 March
Hemsworth and South
Elmsall Express (1988)   20 March
The Independent (1992)   1 February
The Independent (1992)   28 March
The Independent (1992)   13 June
The Miner (1992)   August
Nottingham Evening Post (1990)   8 May
The Times (1960)   22 January
The Times (1973)   31 May
The Times (1975)   'Yorkshire miners
                                             may take over
                                             money-losing
                                             plant to save
                                             jobs and try to
                                             make a profit',
                                             1 November
The Times (1976)   13 February
The Times (1976)   20 February
The Times (1976)   1 March
The Times (1976)   9 March
The Times (1978)   9 June
The Times (1982)   5 March
The Times (1986)   4 June
The Times (1989)   29 July
The Times (1989)   23 August
Yorkshire Post (1992)   'EC deal could
                                             save 20 pits',
                                             13 July

Books, Articles and other Documents

ACOM Secretariat (1991), *Europe's Coalfields, Problems, prospects, policies.* A Report by the ACOM Secretariat, Barnsley, March.

Adeney, M and Lloyd, J (1986), *The Miners' Strike 1984-5.* Loss without Limit, Routledge and Kegan Paul, London.

Allen, J (1988) 'Towards a post-industrial economy?' in Allen, J and Massey, D (eds). *The Economy in Question.* Sage Publications in association with the Open University, London.

Allen, V L (1981), *The Militancy of British Miners.* The Moor Press, Shipley.

Alt, J E and Chrystal, K A (1983), *PoliticalEconomics.* Wheatsheaf, Brighton.

Asda Group plc (1990), *Reports and Accounts 1990.*

Atkinson, M and Coleman, W D (1992), 'Policy Networks, Policy Communities and the Problems of Governance', *Governance: An International Journal of Policy and Administration*, Vol 5, No 2, April.

Bachrach, P and Baratz, M S (1962), 'Two Faces of Power', *American Political Science Review*, 56.

Bacon, R and Eltis, W (1978), *Britain's Economic Problem: Too Few Producers.* Second Edition, Macmillan, London.

Barnsley Development Office (1992), Written Communication to Author, 8 January, Barnsley Metropolitan Borough Council.

Barnsley Metropolitan Borough Council (undated, circa 1984), *Coal Mining and Barnsley.* A Study of Employment Prospects.

Barnsley Metropolitan Borough Council (undated, circa 1991), *The Barnsley Partnership.* A joint venture between Barnsley MBC and Costain Group plc. Publicity brochure.

BBIC (undated, a), *Technical Action Line.* A Service from BBIC. Information Leaflet, Barnsley.

BBIC (undated, b, circa 1991), *Barnsley Business and Innovation Centre.* Publicity and information booklet.

BBIC (undated, c, circa 1991), *Insight into the BBIC*, Newsletter published by Barnsley Business and Innovation Centre.

BBIC (undated, d, circa 1991), *SWORD Innovation*. Publicity leaflet published by Barnsley Business and Innovation Centre.

BBIC (Autumn 1991), *Insight into BBIC*. Newsletter published by Barnsley Business and Innovation Centre.

Beacham, A (1958), 'The Coal Industry' in Burn, D (ed), *The Structure of British Industry: A Symposium*. 77, Cambridge University Press, Cambridge.

Benn, T (1990), *Conflicts of Interest. Diaries 1977-80*. Edited by Ruth Winstone. Hutchinson, London.

Bennett, R J (1990), 'The Incentives to Capital in the UK Enterprise Zones', *Applied Economics*, 22 March.

Benyon, J (1985), 'Going Through the Motions: The Political Agenda, the 1981 Riots and the Scarman Inquiry'. *Parliamentary Affairs*, Vol 38, no 4, Autumn.

Berry, T Capps, T, Cooper, D, Hopper, T and Lowe, T (1985), 'NCB Accounts - a Mine of Mis-Information?', *Accountancy*, 96, January.

Benyon, H (ed) (1985), *Digging Deeper*. Issues in the Miners' Strike. Verso, London.

Benyon, H, Hudson, R, Sadler, D (1991), *A Tale of Two Industries*. The Contraction of Coal and Steel in the North East of England. Open University Press, London.

Birch, D L (1987), *Job Creation in America*. The Free Press (Macmillan), New York.

Blackaby, F T (ed) (1979), *De-industrialisation*, Heinemann, London.

Boulding, P (1989), *State Policies and Industrial Change: Reindustrialisation Programmes in British Steel Closure Areas*. Unpublished PhD Thesis. University of Durham.

British Coal (1990), Written Communication to Author, 16 August.

British Coal Corporation (1990), *Report and Accounts 1989/90*.

255

British Coal Enterprise (1990 a), *Annual Review 1989/90*. Eastwood, Nottinghamshire.
British Coal Enterprise (1990 b), *5 Years On*, A Review Past, Present and Future. London.
Brown, William H, (1991 a), *Northern Region Property Review*.
Brown, William H, (1991 b), *Southern Region Property Review*
Brown, M and Madge, N (1982), *Despite the Welfare State*. A Report on the SSRC/DHSS Programme of Research into Transmitted Deprivation, Heinemann Educational Books, London.
Burch, M and Wood, B, (1983), *Public Policy in Britain*. Basil Blackwell, Oxford.
Burch, M and Wood, B (1990), *Public Policy in Britain*. Second Edition, Basil Blackwell, Oxford.
Burton, J (1983), *Picking Losers?* The Political Economy of Industrial Policy. Hobart Paper 99. Institute of Economic Affairs, London.
Butcher, H, Law, I G, Leach, R and Mullard, M (1990), *Local Government and Thatcherism*. Routeledge, London.
Butler, D and Stokes, D (1969), *Political Change in Britain: The Evolution of Electoral Choice*. Macmillan, London.
Cairncross, A, (1979), 'What is De-Industrialisation?' in F Blackaby (ed), *De-Industrialisation*. Heinemann, London.
Callinicos, A and Simons, M (1985), *The Great Strike*. The Miners' Strike of 1984-5 and its Lessons. A Socialist Worker Publication.
Campbell, M, (1990), 'Employment and the Economy in the 1980s and Beyond', in M Campbell, (ed), *Local Economic Policy*. Cassell Educational, London.
Carley, M, (1980), *Rational Techniques in Policy Analysis*. Policy Studies Institute. Heinemann Educational Books, London.
CCC News (1990), The Newsletter of the Coalfield Communities Campaign, Autumn.
The Centre for Employment Initiatives (1986), *Coalfields Resource and Enterprise Centres*. Phase I and Phase II Reports, January and February, Liverpool.

Chandler, J (1991), *Local Government Today*. Manchester University Press, Manchester.

City of Wakefield Metropolitan District Council (CWMDC), (1981), *Yorkshire's Zones, Developer's Guide*. August.

Coal News (1991), *A Centre for Future Attraction*, No 363, British Coal, Eastwood, Nottingham.

Coalfield Communities Campaign (1986), 'The Written Evidence of the Coalfield Communities Campaign. Presented to the House of Commons Energy Committee During their Investigation Into The Coal Industry in February 1986. Prepared by the Secretariat of the Coalfield Communities Campaign' in *Working Papers*, Vol 4, Coalfield Communities in Campaign, Barnsley.

Coalfield Communities Campaign (undated, circa April 1991), *Coalfield Communities Campaign*, Information and Publicity Booklet, Barnsley.

Coates, D and Hillard, J (ed) (1986), *The Economic Decline of Modern Britain*. The Debate between Left and Right. Harvester Wheatsheaf, Hemel Hempstead.

Coates, K (1992), 'Case for Keeping Pits Open?' Letter to *The Guardian*, 25 September.

Cobb, R, Ross, J K, and Ross, M H (1976), 'Agenda Building as a Comparative Process', *American Political Science Review*, Vol LXX, No 1, p 126-138.

Commission of the European Communities (undated, circa 1989), 'Information Note on the Verification of the Additionality Principle', Letter from the Directorate-Generale for Economic and Financial Affairs to Governments of Member States.

Coopers and Lybrand Deloitte and Sheffield City Polytechnic (1990), *The Dearne Valley Initiative, Economic Study and Business Plan*. July.

Coulson, A (1990), 'Evaluating Local Economic Policy'. in Campbell, M (ed), *Local Economic Policy*. Cassell, London.

Coulter J, Miller S, Walker M (1984), *State of Seige*, Politics and Policing of the

Coalfields: Miners' Strike 1984. Canary Press, London.

Coyne, P (1982), 'Mining Disaster in 1992: 30,000 Lost', *New Statesmen Society*, 16 October, Vol 5.

Craven Tasker (1990), Letter to Author, 19 December.

CWMDC, (1993), *Wakefield's Enterprise Zones. Developer's Guide*, September. City of Wakefield Metropolitan District Council.

CWMDC, (undated, circa 1988/89), *Elmsall Profile*.

CWMDC, (undated, circa 1989), *Wakefield: A Declining Industrial Region*.

CWMDC (1990), Written Communication to Author, 31 August.

Daniels, P (1988), 'Producer Services and the Post-Industrial Space-Economy', in Massey, D and Allen, J (eds), *Uneven Redevelopment*. Cities and Regions in Transition, Hodder and Stoughton in association with the Open University, London.

Dearlove, J and Saunders, P (1984), *Introduction to British Politics*. Analysing a Capital Democracy. Polity, Cambridge.

Department of Employment (1991), *Employment Gazette*, Vol 99, No 2, HMSO, London, p45-100, February.

Department of Energy (1974), *Coal Industry Examination Final Report*. Department of Energy, London.

Department of Energy (1977), *Coal for the Future*. Progress with 'Plan for Coal' and Prospects to the Year 2000. Department of Energy, London.

Department of Trade and Industry (1983), *Regional Industrial Development*. Cmnd 9111, HMSO, London.

Department of Trade and Industry (1988), *DTI - the department for Enterprise*. Cmnd 278, HMSO, London.

Derbyshire County Council (1990), *Structure Plan*, 1 May.

Derbyshire County Council (1991), *North Derbyshire Coalfield Partnership Project*. Progress on 1990/91 Programme at 31 March 1991.

Dickson, T and Judge, E (eds) (1987), *The Politics of Industrial Closure*. Macmillan Press, London.

Doncaster Metropolitan Borough Council (1987), *Minutes of Council Meeting*, 5 October.

Doncaster Metropolitan Borough Council (1989), *Minutes of Council Meetings*, 9 October-20 November.

Doncaster Metropolitan Borough Council (1990), *Minutes of Council Meetings*, 31 July, 9 July-20 August.

Doncaster Task Force (1990), The Newspaper of The Government Inner City Initiative in Doncaster, Issue No 3, September.

Edwards J and Batley R (1978), *The Politics of Positive Discrimination*. An Evaluation of the Urban Programme 1967-77. Tavistock Publications, London.

English Estates (1990). Written Communication to Author, 28 September.

Extel Financial Unquoted Companies Service (1991) *Frigoscandia Ltd*. Annual Card, November.

Field, J (1986), 'New Mining Developments and Their Impact on the Community', in Field J (ed), *Coal Mining and the Community*. Perspectives from the Warwickshire Coalfield Debate. University of Warwick, Open Studies Paper Number 1.

Fothergill, S and Gudgin, G (1985), 'Job Prospects in the Coalfields', in *Working Papers*, Vol 1, Coalfield Communities Campaign, Barnsley, p9-17.

Fothergill, S and Witt, S. (1990), *The Privatisation of British Coal: An Assessment of its impact on mining areas*. Coalfield Communities Campaign, Barnsley.

Frankel, B (1987), (ed), *The Post-Industrial Utopians*. Polity Press, London.

Gamble, A (1981), *Britain in Decline: Economic Policy, Political Strategy and the British State*. Macmillan, London.

Gladstone, B and Turner, R (1992), 'Thurcroft's Tale', *New Statesman and Society*, Vol 5, 16 October.

Gladstone, B, Geddes, M and Bennington, J (1992), 'Changing the Rules of the Games: Local Authority Networks and Sectoral Industrial Policy', *Local Work*, Vol 38, September.

Glyn, A (undated), *The Economic Case Against Pit Closures*. National Union of Mineworkers. Sheffield.

Goodman, G (1985), *The Miners' Strike*. Pluto Press, London.

Gorz, A (1985), *Paths to Paradise*. On The Liberation from Work. Pluto Press, London.

Grant, W (1982). *The Political Economy of Industrial Policy*. Butterworths, London.

Grant, W (1988), 'Why Governments Consult Pressure Groups', *Social Studies Review*, Vol 3, No 4, March.

Grant, W (1990), 'Insider and Outsider Pressure Groups', *Social Studies Review*, Vol 5, No 3, January.

Grant W and Nath, S (1984), *The Politics of Economic Policy Making*. Basil Blackwell, Oxford.

Grant, W P, Paterson, W and Whitson, C (1988), *Government and the Chemical Industry: A Comparative Study of Britain and West Germany*. The Clarendon Press, Oxford.

Guide to the Coalfields (1968, 1969, 1971), Colliery Guardian, London.

Guide to the Coalfields (1979, 1986, 1990), Colliery Guardian, Redhill, Surrey.

Gulliver, S (1984), 'The Area Projects of the Scottish Development Agency', *Town Planning Review*, Vol 55.

Hall, P (1983), 'Enterprise Zones: A Justification', *International Journal of Urban and Regional Research*, 6, pp 416–421, September.

Hall, P H (1986), 'The Theory and Practice of Innovation Policy: An Overview', in Peter Hall (ed), *Technology, Innovation and Economic Policy*. Philip Allen Publishers, Oxford.

Hall, P (1988), 'The Geography of the Fifth Kondratieff', in Massey, D and Allen, J (eds) *Uneven Redevelopment. Cities and Regions in Transition.* Hodder and Stoughton in association with the Open University, London.

Hall, P, Land, H, Parker, R and Webb, A (1975), *Change, Choice and Conflict in Social Policy.* Heinemann Educational, London.

Hall, T (1981), *King Coal.* Miners, Coal and Britain's Industrial Future, Penguin, Harmondsworth.

Ham, C and Hill, M (1984), *The Policy Process in the Modern Capitalist State*, Wheatsheaf, Brighton.

Handy, C (1984), *The Future of Work - A Guide to a Changing Society*, Basil Blackwell, Oxford.

Harris, L (1988), 'The UK Economy at a Crossroads', in Allen, J and Massey, D (ed), *The Economy in Question.* Sage Publications in association with the Open University, London.

Haughton, G and Roberts, P (1990), 'Government Urban Economic Policy 1979-89: Problems and Potential', in Campbell, M (ed), *Local Economic Policy.* Cassel Educational, London.

Heclo, H (1972), ' Review Article: Policy Analysis', *British Journal of Political Science*, 2.

Heclo, H and Wildavsky, A (1981), *The Private Government of Public Money.* Community and Policy Inside British Politics. Second Edition. Macmillan, London.

Heughan, H E (1953), *Pit Closures at Shotts and the Migration of Miners.* University of Edinburgh Social Sciences Research Centre, Monograph No 1.

Hirschman, A O (1970), *Exit, Voice and Loyalty.* Responses to Decline in Firms, Organisations and States. Harvard University Press, London.

Hirst, P and Zeitlin, J (eds), (1989), *Reversing Industrial Decline?* Industrial Structure and Policy in Britain and Her Competitors. Berg Publishers Limited, London.

HMSO (1988), *Action for Cities*. London. Hogwood, B W (1987), *Recent Developments in British Regional Policy*. Strathclyde Papers on Government and Politics.

Hogwood, B W and Gunn, L A (1984), *Policy Analysis for the Real World*. Oxford University Press, Oxford.

Howell, D (1989), *The Politics of the NUM*. A Lancashire View. Manchester University Press, Manchester.

Hudson, R and Sadler, D (1985), 'Coal and Dole: Employment Policies in the Coalfields', in Beynon, H (ed), *Digging Deeper*. Issues in the Miners' Strike. Verso, London.

Hudson, R and Sadler D (1987), 'National Policies and Local Economic Initiatives: Evaluating the Effectiveness of UK Coal and Steel Closure Area Reindustrialisation Measures', *Local Economy*, Vol 2, No 2, p107-114.

Hudson, R and Sadler, D (1992), 'New Jobs for Old? Reindustrialisation Policies in Derwentside in the 1980s', *Local Economy*, Vol 6, No 4, p316-25. February.

Illich, I (1973), *Tools for Conviviality*. Calder and Boyers, London.

In Business Now (1992), 'Enterprise Initiative Special, No 49, Department of Trade and Industry, London, Spring.

Iron and Steel Statistics Bureau (1986). Country Books, London.

Jenkins, C and Sherman, B (1979), *The Collapse of Work*. Eyre Methuen, London.

Jordan, A G (1990), 'Sub-governments, Policy Communities and Networks, *Journal of Theoretical Politics, 2*.

Jordan, A G and Richardson, J (1987), *British Politics and the Policy Process*. An Arena Approach, Allen and Unwin, London.

Keat, R (1991), 'Introduction: Starship Britain or Universal Enterprise?' in Keat, R and Abercrombie, N. *Enterprise Culture*. Routledge, London.

Kirk, G (ed) (1982), *Schumacher on Energy*. Jonathan Cape, London.

Krieger, J (1984), *Undermining Capitalism*. State Ownership and the Dialectic of Control in the British Coal Industry. Pluto, London.

Lawless, P (1989), *Britain's Inner Cities*. Second Edition. Paul Chapman Publishing Limited, London.

Ledis (1990), 'A European Community Initiative to Encourage the Economic Restructuring of Coal Mining Areas - "RECHAR",' Policy and Resources p9. The Planning Exchange. March.

Leigh, R and North, D (1986), 'Innovation Centres: The Policy Options for Local Authorities', *Local Economy*, number 2, pp 45-56, Summer.

Levie, H, Gregory, D and Lorentzen, N (ed) (1984), *Fighting Closures*, Deindustrialisation and the Trade Unions 1979 1983. Spokesman, Nottingham.

Lewis, J and Townsend, A (ed) (1989), *The North-South Divide*. Regional Change in Britain in the 1980s. Paul Chapman Publishing, London.

Lipsey, R H (1963), *An Introduction to Positive Economics*. Weidenfield and Nicholson, London.

Local Economy (1989), 'MILAN', vol 4, no 2, August. Lukes, S (1974), *Power*. A Radical View. Macmillan Education, Basingstoke.

Macdonald, D F (1976), *The State and the Trade Unions*. Second Edition, Macmillan, London.

MacGregor, I (1986), *The Enemies Within*. The Story of The Miners' Strike, 1984-5. Collins, London.

Mackay, R (1992), 'Regional Inequality, Economic Integration and Automatic Stabilisers', Paper presented to IV World Congress of the Regional Science Association International, Palma, Mallorca, 26-29 May.

MacLachlan, H V (1983), 'Townsend and the Concept of Poverty', *Social Policy and Administration*, 22.

263

Martin, R (1988), 'Industrial Capitalism in Transition: The Contemporary Reorganisation of the British space-economy', in Massey, D and Allen, J, *Uneven Re-Development*. op cit.

Martin, S (1989), 'New Jobs in the Inner City: The Employment Impacts of Projects Assisted Under the Urban Development Grant Programme', *Urban Studies*, Vol 26, No 6, December.

Martin, S (1990), 'City Grants, Urban Development Grants and Urban Regeneration Grants', in Campbell, M (ed), *Local Economic Policy*. Cassell, London.

Massey, D (1983), 'Industrial Restructuring as Class Restructuring: Production Decentralisation and Local Uniqueness', *Regional Studies*, 17.

Mawson, J (ed) (1986), 'Policy Review Section', *Regional Studies*, Vol 18, No 6, p507.

McNulty, D (1987), 'Local Dimension of Closure', in Dickson, T and Judge, D (eds), *The Politics of Industrial Closure*. op cit.

Meegan, R (1988), 'A Crisis of Mass Production?' in Allen, J and Massey, D (eds), *The Economy in Question*. op cit.

Meegan, R (1990), 'Merseyside in Crisis and in Conflict', in Harlow, M, Pickvance, C and Urry, J (1990), *Place, Policy and Politics, Do Localities Matter?* Unwin Hyman, London.

Metcalfe, J S (1986), 'Technological Innovation and the Competitive Process', in Peter Hall (ed), op cit.

Minford, P (1984), 'Banking on the Supply Side', *Management Today*, December.

Minford, P and Kung, P (1988), 'The Costs and Benefits of Coal Pit Closures', in Cooper, D and Hopper, T (eds), *Debating Coal Closures*. Economic Calculation in the Coal Dispute 1984-5. Cambridge University Press, Cambridge.

Monopolies and Mergers Commission (1983), *National Coal Board*. A Report on the Efficiency and Costs in the Development, Production and Supply of Coal by the NCB. Appendices. Volume Two, HMSO, London. Cmnd, 8920, June.

Morgan, K (1985), 'Regional Regeneration in Britain: The "Territorial Imperative" and the Conservative State', *Political Studies*, XXXIII, 4, December. 560-577.

Morgan, K and Sayer, A (1988), 'A "Modern" Industry in a "Mature" Region: The Remaking of Management-Labour Relations', in Massey, D and Allen, J, *Uneven Re-Development*. op cit.

Morgan, W J and Coates, K (undated, circa 1990), *The Nottinghamshire Coalfield and the British Miners' Strike 1984-85*. Occasional Paper, University of Nottingham. Department of Adult Education.

Morris, P (1991), "Freeing the Spirit of Enterprise: the Genesis and Development of the Concept of Enterprise Culture", in Keat, R and Abercrombie, N, op cit.

Municipal Year Book (1990) and the Public Services Directory, Municipal Journal, London.

Municipal Year Book (1991) and the Public Services Directory, Municipal Journal, London.

Municipal Year Book (1992) and the Public Services Directory, Municipal Journal Limited, London.

National Audit Office (1986), *Report by the Comptroller and Auditor General*. Department of the Environment, Scottish Office and Welsh Office: Enterprise Zones. HMSO. London.

National Coal Board (1975), *Report and Accounts 1974/75*. NCB, London.

National Coal Board (1981), *Report and Accounts 1980/81*. NCB, London.

National Coal Board (1983), *Report and Accounts 1982/83*. NCB, London.

National Coal Board (1984), *Report and Accounts 1983/84*. NCB, London.

National Coal Board (1985), *Report and Accounts 1984/85*. NCB, London.

NCB Enterprise (1985), *NCB Enterprise Initiative*. October.

Newton, K and Karran T J (1985), *The Politics of Local Expenditure*. Macmillan, London.

North Derbyshire Coalfield Partnership (1990), *Coalfield Development Strategy and 1991-92 Programme*. Submission to the Rural Development Commission. December.

North Derbyshire Coalfield Partnership (1991), *Coalfield Development Strategy and 1992-93 Programme*. Submission of the Rural Development Commission. October.

Northcott, J with Walling A (1988), *The Impact of Microelectronics Diffusion, Benefits and Problems in British Industry*. Policy Studies Institute, London.

Northern Flagship for Industry and Commerce (undated, circa 1992), published in conjunction with the Department of Trade and Industry by Tweedprint Ltd, 97 Heaton Street, Standish, Near Wigan, WN6 ODA.

O'Donnell, K (1988a), *The Impact of Job Losses in the Coal Mining Industry on Wakefield Metropolitan District* 1981-88. Unpublished paper commissioned by the City of Wakefield Metropolitan District Council.

O'Donnell, K (1988b), 'Pit Closures in the British Coal Industry: A Comparison of the 1960s and 1980s', *International Review of Applied Economics*, Vol 2, Part 1.

O'Dowd, L and Rolston, B (1985), 'Bringing Hong Kong To Belfast? The Case of an Enterprise Zone', *International Journal of Urban and Regional Research*, 9 June.

OECD (1971), *The Conditions for Success in Technological Innovation*, Organisation for Economic Co-Operation and Development, Paris.

OECD (1979), *The Case for Positive Adjustment Policies*. A Compendium of OECD Documents 1978/79. Organisation for Economic Co-Operation and Development, Paris.

OECD (1984), *A Positive Adjustment in Manpower and Social Policies*. Organisation for Economic Co-Operation and Development, Paris.

OECD (1987), *Revitalising Urban Economies*. Organisation for Economic Co-Operation and Development. Paris.

O'Keefe, P, Chadwick, M J, Hill R and Robinson D (1989), *How Green is My Power Station?* Coalfield Communities Campaign, Barnsley.

O'Shaughnessy, T. (1990), *Coal Imports: The Macroeconomic Implications for the United Kingdom*, Coalfield Communities Campaign, Barnsley.

Owen G (1988), *British Coal Enterprise - A First Assessment*, A Coalfield Communities Campaign, Barnsley. P A Cambridge Economic Consultants (1987), *An Evaluation of the Enterprise Zone Experiment*. Department of the Environment. HMSO, London.

Parker, J E S (1978), *The Economics of Innovation*. The National and Multinational Enterprise in Technological Change. Second Edition. Longman. London.

Parkinson, E (1985), 'Coalfield Communities and their Environment', *Working Papers*, Volume 2, Coalfield Communities Campaign, Barnsley, p1-9.

Parkinson, M and Duffy J (1984), 'The Minister for Merseyside and the Task Force', *Parliamentary Affairs*, 37, p76-96'.

Peat Marwick McLintock (1991), *Liquidation Highest for Twenty One Years*. Press Information, 20 May.

Platt S and Lewis J (1988), 'Thatcher's Blueprint for the Inner Cities', *New Society*, 11 March 83.

Pollard, S (1982), *The Wasting of the British Economy*. Croom Helm, London.

Polsby, N W (1963), *Community Power and Political Theory*. Yale University Press, New Haven.

Priscott, D (1981), 'Interview with Arthur Scargill', *Marxism Today*, p 5-10. April.

Redbrook/Woolley Campaign Group (undated), *Keep Redbrook and Woolley Collieries Open*. Campaign Publicity Leaflet.

Rhodes, R A W (1986), *Beyond Westminster and Whitehall*. Unwin Hyman, London.

Ridley, N (1991), *My Style of Government*. The Thatcher Years. Hutchinson, London.

Roberts, P and Green, H (1990), *Room to Grow?* Land and Property for Economic Development in Coal Areas. Special Report, Coalfield Communities Campaign, Barnsley.

Robinson, W (1984), 'How Large a Coal Industry?', *Economic Outlook* 1984-1988. The London Business School with Gower Publishing. December.

Romaya, S M and Alden, J D (1987), *The Valleys Initiative, Impact Evaluation Study.* Preliminary Report. Welsh Office. April.

Romaya, S M and Alden, J D (1988), *The Valleys Initiative, Impact Evaluation Study.* Final Report. Welsh Office. July.

Rothwell, R (1986), '*Reindustrialisation, Innovation and Public Policy*', in Peter Hall (ed) op cit.

Rowley, C K (1973), *Antitrust and Economic Efficiency.* Macmillan, London.

Rural Development Commission (1988), *Annual Report 1987/88.*

Sable, C F (1989), 'Flexible Specialisation and the Re-emergence of Regional Economies', in Hirst P and Zeitlin, J (ed) (1989) op cit.

Samuel, R, Bloomfield, B and Boanas, G (1986), *The Enemy Within.* Pit Villages and the Miners' Strike of 1984-5. Routledge and Kegal Paul, London.

Scammel, M (1986), *The Enemy Within. Government and the Miners' Strike.* Strathclyde Papers on Government and Politics.

Scargill, A (1975), 'The New Unionism', *New Left Review*, 92, July-August.

Scargill, A and Kahn, P (1980), *The Myth of Workers' Control*, Occasional Paper, University of Leeds and Nottingham.

Singh, A (1977), 'UK Industry and the World Economy: A Case of Deindustrialisation?' *Cambridge Journal of Economics*, Vol 1, p113-36. June.

Smith, M (1989), Land Nationalisation and Agricultural Policy Community', *Public Policy and Administration*, Vol 4, No 3, Winter.

Solesbury, W (1976), 'The Environmental Agenda: An Illustration of How Situations May Become Political Issues and Issues May Demand Responses From Government: Or How They May Not', *Public Administration*, Vol 54, p379-3978. Winter.

Stewart, J D (1958), *British Pressure Groups*. Oxford University Press, Oxford.

Stewart, M (1987), 'Ten Years of Inner Cities Policy', *Town Planning Review*, 58.

Storey, D J (1990), 'Evaluation of Policies and Measures to Create Local Employment, *Urban Studies*, Vol 27, No 5, p669-684.

Stoker, G (1988), *The Policies of Local Government*. Macmillan, London.

Stroetmann, R (1979), 'Innovation in Small and Medium Sized Firms', in Baker, M J (ed), *Industrial Innovation*. Macmillan, London.

Taylor, A (1984), *The Politics of the Yorkshire Miners*. Croom Helm, London.

The Times Guide to the House of Commons (June 1987), published by Times Books, a division of Harper Collins Publishers, London.

Thomas, I C (1992), 'Additionality in the Distribution of European Regional Development Fund Grants to Local Authorities', *Local Economy*, Vol 6, No 4, February.

Thomas, M (1989), Colliery Closure, Entrepreneurship and Industrial Theory', Coalfield Communities Campaign, *Working Papers 5* p23-32, Barnsley.

Turner, R L (1985), 'Post-War Pit Closures: The Politics of Deindustrialisation', *The Political Quarterly*, Vol 56, 2, April-June.

Turner, R L (1988), 'Representation and Power in Britain's One-Party States: The Case of Barnsley' *Teaching Public Administration*, Vol VIII, No 2, Autumn, p24-34.

Turner, R L (1989), *The Politics of Industry*. Christopher Helm, Bromley, Kent.

Turner, R L (1990), 'Mrs Thatcher's "Enterprise Culture". Is It Any Nearer?', *Social Studies Review*, Vol 5, No 3, p120 -122, January.

Turner, R L (1991), 'An Enterprise Zone In A Pit
    Closure Zone: The Politics of Industrial
    Subsidies', Paper Presented to the *Regional
    Science Association 31st European Congress*,
    Lisbon, Portugal, 27-30 August.
Turner, R L (1992a), 'The Role of "New
    Entrepreneurship" and "Innovation" in
    Regenerating the Coalfields: Evidence From
    South Yorkshire and North Derbyshire',
    Paper Presented to IV World Congress of the
    R e g i o n a l   S c i e n c e   A s s o c i a t i o n
    International, Palma de Mallorca, 26-29
    May.
Turner, R L (1992b), 'A Task Force in a Locality
    of Declining Coal Mining: The Case of
    Doncaster', *The East Midland Geographer*,
    Volume 15, Part 1, p16-29.
Turner, R L (1992c), 'British Coal Enterprise:
    Bringing the "Enterprise Culture" to a
    Deindustrialised Local Economy?' *Local
    Economy*, Volume 7, Number 2, p4-8. May.
Waller, R (1983), *The Almanac of British
    Politics*. Croom Helm, London.
Welsh Office (undated), *The Valleys. A
    Partnership with the People*.
Welsh Office Information Division (1988), *A
    Programme for the People*.
West Lothian Business Information Bureau (1990),
    Development and Change. Newsletter, Issue
    10, January.
Wiener, M (1981), *English Culture and the
    Decline of the Industrial Spirit* 1850-1980.
    Cambridge University Press. Cambridge.
Wilks, S (1984), *Industrial Policy and the Motor
    Industry*. Manchester University Press.
    Manchester.
Wilks, S and Wright, M (1987), 'Conclusion:
    Comparing Government - Industry Relations:
    States, Sectors and Networks', in Wilks, S
    and Wright, M (eds), *Comparative Government
    - Industry Relations*. Oxford University
    Press, Oxford.
Williams, B (1986), 'Technical Change and
    Employment', in Peter Hall (ed), op cit.
Williams, K (1983), 'Introduction: Why are the
    British Bad at Manufacturing? in Williams
    K, Williams J and Thomas, D. *Why Are The*

British Bad at Manufacturing? Routledge and Kegan Paul, London.

Williams K, Williams J and Thomas D (1983), *Why Are The British Bad at Manufacturing?* Routledge and Kegan Paul, London.

Winterton, J and Winterton, R. *Coal, Crisis, Conflict*. The 1984-85 Miners' Strike in Yorkshire. Manchester University Press, Manchester.

Witt, S (1990), *When The Pit Closes* - The Employment Experiences of Redundant Miners. A Special Report produced by the Coalfield Communities Campaign, Barnsley, 52.

Wright, M (1988), 'Policy Community, Policy Network and Comparative Industrial Policies', *Political Studies*, Vol 36.

Yellow Pages (1992), Sheffield 1992/93. British Telecommunications plc, Reading.

Young, S (1985), 'The Scope for Locally Based Social and Economic Initiatives in the Coalfield Communities', *Working Papers*, Vol 2, Coalfield Communities Campaign, p17-23, Barnsley.